4th International Conference on Informatics, Engineering, Science and Technology (INCITEST 2021)

Held online due to COVID-19

IOP Conference Series: Materials Science and Engineering
Volume 1158

Bandung, Indonesia
10 March 2021

ISBN: 978-1-7138-3426-7
ISSN: 1757-8981

Printed from e-media with permission by:

Curran Associates, Inc.
57 Morehouse Lane
Red Hook, NY 12571

Some format issues inherent in the e-media version may also appear in this print version.

No changes have been made to the content of these proceedings. There may be changes to pagination and minor adjustments for aesthetics.

Printed with permission by Curran Associates, Inc. (2021)

Additional copies of this publication are available from:

Curran Associates, Inc.
57 Morehouse Lane
Red Hook, NY 12571 USA
Phone: 845-758-0400
Fax: 845-758-2633
Email: curran@proceedings.com
Web: www.proceedings.com

TABLE OF CONTENTS

PROCESSING, ENGINEERING AND APPLICATIONS

COMPUTATION AND SIMULATION

IOP Conf. Series: Materials Science and Engineering 1158 (2021) 011001 doi:10.1088/1757-899X/1158/1/011001

PREFACE

We are honored to introduce the Proceedings of the 4th International Conference on Informatics, Engineering, Science, and Technology (INCITEST 2021). With its main theme on "A Future Road Map, Encouraging Optimism and Technology to Face Post Covid-19 Era", this event is hoped to serve as a platform of gathering for anyone interested in exploring optimism, potential solutions, and answering issues and challenges in the era Covid-19 Pandemic". Unlike the 1st and 2nd INCITEST, the 3rd and 4th INCITEST was carried out in the virtual conference due to the impact of COVID-19. Universitas Komputer Indonesia (UNIKOM) held this online conference on 10 March 2021 to gather experts and scholars worldwide to continue disseminating the latest advanced research in the field of Informatics, Engineering, Science, and Technology and developing the academic exchange among researchers.

Following the success of INCITEST 2018, 2019, and 2020, the enthusiastic responses to the call-for-papers in the fourth INCITEST were increasing. More than 150 papers were submitted to the organizing committee from both local and foreign participants. A peer-review process has been conducted to check paper originality and quality, resulting in 120 accepted papers being presented. The papers have been presented via Zoom in 10 minutes for each paper.

It is an honor for us to have four renowned experts as our keynote speakers : Prof. Minoru Terano and Prof. Hiroyuki Iida from Japan Advanced Institute of Science and Technology (JAIST), Dr. Siva Kumar Sivaneshan from Taylor's University, and Dr. Salmon Priadji Martana from UNIKOM. The keynote speech is presented in the plenary session in online format (via Zoom) and lively broadcasted on UNIKOM Youtube official channel. Each keynote presented their address in 45 minutes duration, including the discussion session. After completing the plenary session, the schedule continued with a parallel session held in an online format. The participants were divided into several classes, which a moderator chaired. Each presenter should present their paper within 5 minutes via Zoom. Five minutes after the presentation will be given to each of them for the discussion session.

We believe that the outcome of the conference will be fruitful with the above arrangement. The conference's success is also due to the hard work of all involved parties. Therefore, we would like to express appreciation to all committee members, supporters, sponsors, and participants for an outstanding contribution to the conference's success. High appreciation goes to all of the reviewers who helped us maintain the quality of manuscripts included in the Proceedings published by IOP. We also express our sincere thanks to our Rector, Prof. Dr. Ir. Eddy Soeryanto Soegoto, MT, for his support at the conference. Special thanks are also conveyed to our prestigious publisher partner, IOP, and their publication team for their hard work on our publication.

Finally, this conference series's continuing success can be one indicator that we have through our right pathway to enforce optimism and bring a better roadmap after the pandemic. We hope that the 4th INCITEST will get fruitful outcomes and give the participants a great experience at the conference.

Thank you

Best Regards,

Dr. Lia Warlina
The Chief of the Conference

IOP Conf. Series: Materials Science and Engineering 1158 (2021) 011001 doi:10.1088/1757-899X/1158/1/011001

LIST OF COMMITTEES OF INCITEST 2021

Organizing Committee:

1. Dr. Lia Warlina.
2. Dr. Poni Sukaesih Kurniati, S.IP., M.Si.
3. Bobi Kurniawan, S.T., M.Kom
4. Senny Luckyardi, S.P., MM
5. Ferry Stephanus Suwita, S.Kom., MT.

Advisory Board & Scientific Committee:

1. Prof. Ryo Takahashi - Sendai University, Japan
2. PD Dr. Andreas Ufen - German Institute of Global and Area Studies, Germany
3. Prof. Hiroyuki Iida - Japan Advanced Institute of Science and Technology
4. Prof. Shang Chao Hung - Fuzhou Polytechnic, China
5. Prof. Nursuriati Jamil - University Technology MARA, Malaysia
6. Prof. Hideaki Kanai - Japan Advanced Institute of Science and Technology
7. Dr. Eng. Farid Triawan - Tokyo Institute of Technology, Japan
8. Prof. Rongtau Hou - Nanjing University of Information Science and Technology, China
9. Prof. Tinghuai Ma - Nanjing University of Information Science and Technology, China
10. Assoc. Prof. Bing Han - Xidian University, China
11. Assoc Prof M. Roil Bilad - Universiti Teknologi Petronas, Malaysia
12. Assoc Prof. Zulfan Adi - Universiti Teknologi Petronas, Malaysia
13. Assoc. Prof. Dr. Masnizah Mohd - University Kebangsaan Malaysia
14. Dr. Nordin Abu-Bakar - Universiti Teknologi Mara (UiTM) Malaysia
15. Prof. Dr. Ir. H Eddy Soeryanto Soegoto, M.T - Universitas Komputer Indonesia
16. Dr. Ir. Herman S.,MBA. - Universitas Komputer Indonesia
17. Dr. Jeffry Handoko Putra, MT. - Universitas Komputer Indonesia
18. Dr. Andi Harapan, M.T - Universitas Komputer Indonesia
19. Dr. Dhini Dewiyanti Tantarto, M.T - Universitas Komputer Indonesia
20. Dr. Henny, ST.,M.T - Universitas Komputer Indonesia
21. Dr. Salmon Priadji Martana, ST.,M.T - Universitas Komputer Indonesia
22. Dr. Y. Djoko Setiarto, ST., M.T - Universitas Komputer Indonesia
23. Irfan Dwiguna Sumitra, M.Kom., Ph.D - Universitas Komputer Indonesia
24. Dr. Ade Gafar Abdullah - Universitas Pendidikan Indonesia
25. Dr. Eng. Asep Bayu Dani Nandiyanto - Universitas Pendidikan Indonesia
26. Syeilendra Pramuditya, PhD - Institut Teknologi Bandung, Indonesia
27. Prof. Dr. Norita Md Norwawi - Universiti Sains Islam Malaysia
28. Prof. Dr. Setyawan Purnama, M.Si - Universitas Gadjah Mada
29. Dr. Sandra Sukmaning Adji - Universitas Terbuka

IOP Conf. Series: Materials Science and Engineering 1158 (2021) 011002 doi:10.1088/1757-899X/1158/1/011002

Peer review declaration

Papers published in this volume of *IOP Conference Series: Materials Science and Engineering* have been peer reviewed through processes administered by the Editors. Reviews were conducted by expert referees to the professional and scientific standards expected of a proceedings journal published by IOP Publishing.

All papers are fully peer-reviewed. We only submit papers that have been reviewed and approved by highly qualified researchers with expertise in a field appropriate for the paper. We used single blind peer-reviewing process. The following are the detail procedures of the flow for the paper submission :

1. Abstract Submission Deadline: **28 November 2020**
2. Abstract check and selection (by reviewer and editor): **28 November- 10 December 2020.** In this step, the reviewer will reject the manuscript that out of conference scope. In addition, for the accepted abstract the reviewer will give the comment to revise the abstract in the full paper (if there any issues to be revised).
3. Notification of Abstract Acceptance: **12 December 2020**
4. Full Paper Submission Deadline: **30 December 2020**
5. Paper Check and Selection (by reviewer and editors stage 1): **30 December 2020 - 7 January 2021.** Reviewer have a right to directly accept, reject, or review. Prior to further processing steps, plagiarism check using turtitin is applied for each papers. Reviewer will give comment on the writing scructure, content, language, template, figure, table and chart, reference, and citation style.
6. Notification paper review result (stage 1 review): **8 January 2021**
7. Revised paper submission deadline (from stage 1 review): **14 January 2021**
8. Paper Check and Selection (by reviewer and editors stage 2): **15 January - 18 January 2021.**
9. Notification of Paper Review Result (stage 2) : **19 January 2021**
10. Final Revised Paper Submission Deadline: **23 January 2021**
11. Notification of Paper Acceptance: **13 February 2021**
12. Announcement of Presentation Schedule: **25 February 2021**
13. Conference Date: **10 March 2021**

The reviewer considers the feedback provided by the peer reviewers and arrives at a decision. The following are the most common decisions:

1. accept without any changes (acceptance): the paper will be presented in the conference and will be submitted to the publisher in its original form
2. accept with minor revisions (acceptance): the paper will be submitted and reviewer asks the author to make small corrections
3. accept after major revisions (conditional acceptance): the paper will be submitted to publisher and the authors should make the changes suggested by the reviewers and/or editors
4. reject the paper (outright rejection): the paper will be not submitted to the publisher.

- Conference submission management system:

The papers were submitted to the conference system, handled by the editor. The handling editor managed the correspondence between reviewers and authors.

- Number of submissions received: 120
- Number of submissions sent for review: 120
- Number of submissions accepted: 74
- Acceptance Rate (Number of Submissions Accepted / Number of Submissions Received X 100): 61.7%

- Average number of reviews per paper: 2
- Total number of reviewers involved: 6 reviewers
- Contact person for queries:

senny@unikom.ac.id
senny@email.unikom.ac.id

Material Selection for Vertical Axis Tidal Current Turbine using Multiple Attribute Decision Making (MADM)

S Musabikha[1*], H Pratikno[1], I K A P Utama[2], Mukhtasor[1]
[1] Building WA, Department of Ocean Engineering, Faculty of Marine Technology, Institut Teknologi Sepuluh Nopember (ITS), Surabaya, 60111 Indonesia
[2] Building W, Department of Naval Architecture and Shipbuilding Engineering, Faculty of Marine Technology, Institut Teknologi Sepuluh Nopember (ITS), Surabaya, 60111 Indonesia

*siti15@mhs.na.its.ac.id

Abstract. Marine renewable energy technology has attracted many stakeholders to develop due to its enormous resources and future potential advantages. One of them is the vertical axis tidal current turbine technology, which converts the kinetic energy of flowing seawater to be electrical energy. To operate the vertical axis tidal current turbine, especially its metallic components of the supporting structure, it is required to endure the aggressive and corrosive marine environment. Thus, deciding the good and suitable material for this application is very important in order to reduce the risk and hazards from the sea environment. The purpose of this paper is to decide the best metal material for the support structure of the vertical axis tidal current turbine. Therefore, the material candidate and their properties were firstly reviewed and discussed. Then, the material selection was processed through Multiple Attribute Decision Making (MADM) and formulated via the method of a Technique for Order Preference by Similarity to the Ideal Solution (TOPSIS). The result of material selection was carbon steel AISI 1045 as the selected material instead of other candidates such as carbon steel AISI 1018, stainless steel, and nickel aluminium bronze. Thus, it is expected to perform well to be used for the support structure and endure the marine hazards during the vertical axis tidal current turbine application.

1. Introduction

Ocean energy has the potential to have an important role in the world's future energy system [1]. Of all ocean energy technologies, tidal current energy is an eco-friendly and abundant energy source by converting it to electricity [2,3]. Tidal current energy also has distinct advantages of high predictability and regularity, which makes the exploitation of tidal current energy more attractive [4,5]. One of tidal current energy technology types is Enermar's Kobold Turbine that is a full-scale prototype of vertical axis tidal current turbine deployed in Messina Strait, Italy [6]. The main component of vertical axis tidal current turbine consists of a blade, shaft, and connecting arm. While the vertical axis tidal current turbine technologies operate in the sea environment, there are challenges that are common to all. The nature of marine renewable energy technologies means that the greatest challenge is the operation in the marine environment itself [7]. Marine renewable energy devices are exposed to highly dynamic, harsh marine environments, and complex stresses, i.e.: corrosive stresses, mechanical stresses, and biological stresses [8]. Therefore, it is important to provide materials that can meet the vertical turbine design criteria, withstand those challenges, and ensure the survivability of the devices.

Several material selections have been done to choose suitable material for ocean energy technology. Thakker et al. [9] studied a new approach for material selection strategy of wave energy extraction turbine blade material. It resulted that the GFRP was the optimum material and followed by titanium alloys in the second-ranking. Shiekh-Elsouk et al. [10] also reviewed the material selection for tidal turbine blades and discussed the parameters that affect the performance of the tidal turbine. They considered several materials such as carbon steel, stainless steel, aluminum, titanium, and composite and concluded that the composite was the most suitable material for a tidal current turbine blade. The composite material is usually used for the prime mover or the moving parts (turbine blade) of ocean energy [11], thus it can affect the performance efficiency of the turbine blades. Besides, several investigations of composites have been done to obtain the optimum design of tidal turbine blades such as Grogan et al. [12], Davies et al. [13], and Wang et al. [14]. In the other side, the supporting structure of the ocean energy device requires more materials in great quantities compared to the turbine blade [15]. Moreover, the information on the material selection for the support structure of the vertical axis tidal current turbine is still limited.

Therefore, this paper aims to choose the suitable metal material that will be used in the vertical axis tidal current turbine, especially the supporting structure. First, the material candidate and their properties will be reviewed and discussed. Followed by the process of material selection and it will be performed by using multiple attribute decision making (MADM). Lastly, the material selection of the vertical axis tidal current turbine will be concluded.

2. Methods

The decision making is performed using Multiple Criteria Decision Making (MCDM) and one of its types is Multiple Attribute Decision Making (MADM). The MADM is the selection of an alternative from a set of options which is based on prioritized attributes of the options [16]. Several types of methods are formulated based on MADM, and one of them is the Technique for Order Preference by Similarity to the Ideal Solution (TOPSIS) to find the best alternative.

With using the TOPSIS method, the chosen alternative will be near to the ideal solution and it will be far from the negative-ideal solution. There are several steps to conduct the TOPSIS method [17]:

1. Build the decision matrix of the alternatives after reviewing several material candidates, and construct the normalized decision matrix. The criteria are divided by total outcomes vector (the Euclidean length of a vector). So that the elements r_{ij} of the normalized decision matrix R are:

$$r_{ij} = \frac{x_{ij}}{\sqrt{\sum_{i=1}^{m} x_{ij}^2}} \tag{1}$$

2. Construct the weighted normalized decision matrix by setting the value weight $W = (w_1, w_2, \dots . w_n)$, so the weighted normalized matrix V:

$$V = W \times r_{ij} = \begin{bmatrix} w_1 r_{11} & \cdots & w_n r_{1n} \\ \vdots & \ddots & \vdots \\ w_1 r_{m1} & \cdots & w_n r_{mn} \end{bmatrix} \tag{2}$$

3. Determine the value of the ideal solution and negative-ideal solution. The ideal solution V^+ and negative-ideal solutions V^-:

$$V^+ = \{(\max V_{ij} |\, j \in J_1), (\min V_{ij} |\, j \in J_2), i = 1,2,3, \dots, m\} = \{v_1^+, v_2^+, \dots, v_n^+\} \tag{3}$$

$$V^- = \{(\min V_{ij} |\, j \in J_1), (\max V_{ij} |\, j \in J_2), i = 1,2,3, \dots, m\} = \{v_1^-, v_2^-, \dots, v_n^-\} \tag{4}$$

Where:

$J_1 = \{j = 1,2,3 \dots n\}$ and j is related with benefit criteria

$J_2 = \{j = 1,2,3 \dots n\}$ and j is related with cost criteria

4. Estimate the separation measure. Then the separation distance of each alternative from the ideal solution and negative-ideal solutions can be attained by the n-dimensional Euclidean distance method. The distance (in a Euclidean sense) S_i^+ of each alternative from the ideal solution is defined:

$$S_i^+ = \left[\sum_{j=1}^{n} \left(V_{ij} - V_{ij}^+\right)^2\right]^{0,5} \text{ where i=1,2,3...., m.} \tag{5}$$

And the distance from the negative-ideal solution is defined:

$$S_i^- = \left[\sum_{j=1}^{n}\left(V_{ij} - V_{ij}^-\right)^2\right]^{0,5} \text{ where i=1,2,3…., m.} \tag{6}$$

5. Calculate the value or relative closeness to the ideal solution of an alternative:

$$P_i = \frac{S_i^-}{S_i^+ + S_i^-} \text{ where } 0<P_i<1 \text{ and i=1,2,3…m.} \tag{7}$$

6. Rank the order of the results according to the P_i. Hence, the best alternative can be obtained.

3. Results and Discussion

3.1. The Candidate of Material Selection and Their Properties

There are several important factors that should be considered in material selection for the ocean structure: (1) the physical and chemical properties of materials, (2) cost and availability, (3) fabrication facility, and (4) maintenance costs [18]. During the material selection, the physical characteristics of the material are very important to consider. Especially, the yield strength value of material as the first factor that needs to be taken into consideration, and then followed by the value of Young's modulus and ductility. Hence, those factors were used to screen alternative metal materials.

Some metals with the marine grade that is usually applied in marine environment are carbon steel and steel alloy, aluminum, stainless steel, copper, bronze, and brass. Among those metals, carbon steel, stainless steel, and nickel aluminum bronze (NAB) offer good corrosion resistance, hence they are selected to be material candidates used on this research. Carbon steel consists of three types, that is low, medium, and high. For offshore structures, the low or medium carbon steel is usually used, for example, AISI 1018 and AISI 1045. Carbon steel AISI 1018 is mild/low and it is widely used for engineering materials and in ocean applications [19]. AISI 1018 has good strength and ductility, its cost is affordable, and it is easy to cut and to be welded. Carbon steel AISI 1045 is a medium type. It is easy to weld and set up through the machine and can receive heat treatment.

Stainless steel is well known for its general corrosion resistance due to its passive characteristics (oxide film) [20]. It has the composition of chromium, nickel, molybdenum, etc. and its example is SS304, SS316, and SS316L. SS316 and SS316L have better corrosion resistance compared to SS304. The difference between SS316 and SS316L is located on the carbon composition. SS316L has a maximum carbon content of 0.03% compared to SS316 that has a maximum carbon content of 0.08%. "L" represents the abbreviation of "low." Thus, SS316L is easier to weld than SS316. Because of its higher carbon content, SS316 will be more prone to weld decay or welding damage. In other words, the SS316L is more resistant to intergranular corrosion, and their costs also do not differ greatly. SS316L can provide high strength and ductility properties close to carbon steels (S355) [21], which is the main steel grade used in offshore structures. However, SS316L is susceptible to pitting corrosion and crevice corrosion [22,23]. This localized corrosion (pitting and crevice corrosion) emerge because the passive film breaks down influenced by high chloride ion concentration, low dissolved oxygen, and acidity [24].

For NAB, it is one of the copper-based alloys and it has good strength and corrosion resistance that is commonly used in marine applications, especially ship propeller [25,26]. Oldfield and Masters [27] stated that NAB has good pitting corrosion resistance, reasonable erosion-corrosion resistance, comparable resistance to cavitation and fatigue, and immune to stress corrosion cracking (SCC). Its corrosion resistance is originated from a protective oxide film with thickness for about several hundred nanometers [28]. The average erosion-corrosion rate of NAB in fresh unfiltered seawater with velocity over the range of 7.6-30 m/s [29]. For the tidal stream velocity, the peak tidal current velocity ranges about 2-3 m/s which is lower compared to the ship propeller applications [30]. However, the turbine rotating conditions are in quite dynamic and turbulent conditions and so there is a chance that the protective layers are unable to repassivate and local attacks can occur. Therefore, carbon steel AISI 1018 and AISI 1045, stainless steel 316L and nickel aluminum bronze (NAB) C95800 are selected to be candidates used on this material selection.

3.2. Material Selection for Vertical Axis Tidal Current Turbine

The material selection for the vertical axis tidal current turbine was conducted using the TOPSIS method [16,17]. There are four selected material candidates, i.e.: AISI 1018, AISI 1045, NAB C95800, and

SS316L. Table 1 presents the decision matrix of four types of alternative with their properties. For the availability attribute, it was determined by considering the easiness to get the material and the readiness of the vendor to provide materials at any time. The availability value is shown in Table 2. Furthermore, the four material candidate properties related to the application of vertical axis tidal current turbine were evaluated.

Table 1. The decision matrix with four types of alternative from the metallic materials and its properties [19,27,31-33]

	Yield strength, MPa	Young Modulus, GPa	Tensile strength, MPa	Density, g cm^{-3}	Cost, IDR kg^{-1}	Corrosion rate, mm year^{-1}	Availability
AISI 1018	370	205	440	7.87	10400	0.026	4
AISI 1045	450	200	585	7.87	25850	0.02	4
SS 316 L	205	193	485	8	67550	0.5	4
NAB	240	130	620	7.5	486000	0.05	2

Table 2. The values of the material availability

Value	Availability
1	Not Available
2	Less Available
3	Quite Available
4	Available
5	Very Available

To apply the TOPSIS method, the decision matrix was normalized using equation (1). The normalization was performed in order to change the data value on the numeric column into data with the same scale, without disturbing the difference in the value range. It is only necessary if its properties have a different range. The results of the normalized decision value are shown in Table 3. Then, the normalized decision matrix was weighted using equation (2) to demonstrate the importance of material potentialities. The values of weight are: W={5,4,4,2,4,4,5}. The values of weight were based on Jadidi et al. [34] and its results of the weighted normalized decision matrix are shown in Table 4.

Table 3. The normalized decision of the material selection

	Yield strength, MPa	Young Modulus, GPa	Tensile strength, MPa	Density, g cm^{-3}	Cost, IDR kg^{-1}	Corrosion rate, mm year^{-1}	Availability
AISI 1018	0.5584	0.5555	0.4093	0.5037	0.0212	0.0516	0.5547
AISI 1045	0.6792	0.5420	0.5442	0.5037	0.0526	0.0397	0.5547
SS 316 L	0.3094	0.5230	0.4512	0.5120	0.1374	0.9929	0.5547
NAB	0.3622	0.3523	0.5768	0.4800	0.9889	0.0993	0.2774

Table 4. The weighted normalized decision matrix of the material selection

	Yield strength, MPa	Young Modulus, GPa	Tensile strength, MPa	Density, g cm^{-3}	Cost, IDR kg^{-1}	Corrosion rate, mm $year^{-1}$	Availability
AISI 1018	2.7921	2.2221	1.6373	1.0074	0.0846	0.2065	2.7735
AISI 1045	3.3958	2.1679	2.1769	1.0074	0.2104	0.1589	2.7735
SS 316 L	1.5470	2.0920	1.8048	1.0240	0.5498	3.9717	2.7735
NAB	1.8111	1.4091	2.3072	0.9600	3.9555	0.3972	1.3868

Table 5. The matrix of ideal solution V^+ and negative-ideal solution V^- of the material selection

	Yield strength, MPa	Young Modulus, GPa	Tensile strength, MPa	Density, g cm^{-3}	Cost, IDR kg^{-1}	Corrosion rate, mm $year^{-1}$	Availability
V+	3.3958	2.2221	2.3072	1.0240	0.0846	0.1589	2.7735
V-	1.5470	1.4091	1.6373	0.9600	3.9555	3.9717	1.3868

Table 6. The value of separation measure (Si+ and Si–), the value of relative closeness (Pi) to the ideal solution, and the rankings of the best selected material

	Si+	Si-	Pi	Ranking
AISI 1018	0.9031	5.7703	0.8647	2
AISI 1045	0.1897	5.8970	0.9688	1
SS 316 L	4.2943	3.7444	0.4658	3
NAB	4.4878	3.6463	0.4483	4

Subsequently, the value of the ideal solution matrix V^+ and negative-ideal solution matrix V^- (Table 5) were determined and calculated using equations (3) and (4). Then, the distance value (Si+ and Si–) between the separation measure of each alternative with the ideal solution and negative-ideal solution were estimated using equations (5) and (6), and are shown in Table 6. Afterwards, the value of preferences (Pi) or relative closeness to the ideal solution for each alternative were also determined by the equation (7), and the results were also given in Table 6. Finally, the ranking of alternative in this material selection was sorted based on Pi values, and the results can also be obtained in Table 6. The best-selected material is AISI 1045 and this is in agreement with [15] that steel material is selected as the most suitable material for the support structure of tidal current energy devices. Therefore, this material can be expected and able to accommodate the nature of the sea environment and meet the material design requirement for the vertical axis tidal current turbine.

4. Conclusion

The material selection for vertical axis tidal current turbine using the MADM-TOPSIS method has been conducted in this paper. There are several factors to consider in order to select the suitable material: the physical and chemical properties of materials, cost and availability, fabrication facility, and maintenance costs. From the analysis, carbon steel AISI 1045 was chosen as the best material for the vertical axis tidal current turbine, especially for the supporting structure that still needs metal as the main material. Carbon steel AISI 1045 possesses good strength and corrosion resistance, affordable price and it is easy to provide. Therefore, carbon steel AISI 1045 as the chosen material can be expected to be able to meet the design requirement, withstand the marine hazards and avoid catastrophic failure during the operation of vertical axis tidal current turbine.

Acknowledgements

This research was supported by the Education of Master Degree Leading to Doctoral Program for Excellent Graduates (PMDSU) program and funded through the Graduate Research program by Institut Teknologi Sepuluh Nopember, Surabaya (No: 926/PKS/ITS/2020).

References

[1] Magagna, D., and Uihlein, A. 2015. *Int. J. Mar. Energy,* **11**, pp. 84.
[2] Satrio, D., Utama, I. K. A. P., and Mukhtasor. 2018 *J. Mar. Sci. Appl.* **17**, pp. 208.
[3] Junianto, S., Mukhtasor, Prastianto, R. W., and Jo, C. H. 2020. *Oce. Syst. Eng.* **10**, pp. 87.
[4] Li, W., Zhou, H., Liu, H., Lin, Y., and Xu, Q. 2016. *Renew. & Sust. Ener. Reviews* **63**, pp. 414
[5] Musabikha S., Utama, I. K. A. P., and Mukhtasor. 2018. *Proc. Int. Sem. Metal. & Mat. (Jakarta)* p 1
[6] Musabikha, S., Utama, I. K. A. P., and Mukhtasor. 2021. *J. Eng. Sci. Tech., under reviewed*
[7] Mofor, L., Goldsmith, J., and Jones, F. 2014. *Ocean Energy: Technology Readiness, Patents, Deployment Status and Outlook* (Abu Dhabi: International Renewable Energy Agency) p 12
[8] Momber, A. W., Plagemann, P., and Stenzel, V., 2015. *Renew. Energy* **74** 606
[9] Thakker, A., Jarvis J, Buggy M and Sahed A 2008 *Mater. & Desi.* **29** 1973
[10] Shiekh, Elsouk, M. N., Santa, Cruz, A., and Guillou, S. S. 2018. *Proc. 7th Int. Conf. Oce. Ener. (Cherbourg, France),* p. 1.
[11] Tong, C. 2019. *Introduction to Materials for Advanced Energy Systems* (Switzerland: Springer)
[12] Grogan, D. M., Leen, S. B., Kennedy, C. R., and Ó Brádaigh, C. M. 2013. *Renew. Ener.* **57**, pp. 151.
[13] Davies, P., Boisseau, A., Choqueuse, D., Peters, L., Renaud, C., Nickel, R., Thiebaud, F., and Perreux, D. 2010. *Proc. 3th Int. Conf. Oce. Ener. (Bilbao, Spain)* p. 1.
[14] Wang, J., Piechna, J., and Muller, N. 2011. *Arch. Mech. Eng.* **58**, pp. 355
[15] Douglas, C. L., Harrison, G. P., and Chick, J. P. 2008, *Proc. Inst. Mech. Eng. (United Kingdom)*, p. 1
[16] Suresh-Babu, K., Subba-Raju, N. V., Srinivasa-Reddy, M., and Nageswara-Rao, D. 2006. *Proc. 18th Int. MCDM Conf. (Greece)* p. 1.
[17] Hwang, C. L., and Yoon, K. 1981. *Multiple Attribute Decision Making: Methods and Applications* (Heidelberg, Germany: Springer)
[18] Chandrasekaran, S., and Jain, A. K. 2016. *Ocean Structures, Construction, Materials, and Operations* (London: CRC Press)
[19] Toloei, A., Atashi, S., and Pakshir, M. 2013. *Corr. Reviews* **31**, pp. 135.
[20] Burson-Thomas, C. B., and Wood, R. J. K. 2017. *J. Bio Trib. Corr.* **3**, pp. 14.
[21] Baddoo, N. R., and Burgan, B. A. 2001. *Structural Design of Stainless Steel* (Berkshire UK: The Steel Construction Institute), p. 6.
[22] Machuca, L. L., Bailey, S. I., and Gubner, R. 2012. *Corr. Sci.* **64**, pp. 8.
[23] Okeremi, A., and Simon-Thomas, M. 2008. *NACE: Corr. Conf. & Expo (Texas)*, p. 1.
[24] Bhandari, J., Khan, F., Abbassi, R., Garaniya, V., and, Ojeda. R 2015. *J. Loss Prev. Process Ind.* **37**, pp. 39
[25] Wharton, J. A., Barik, R., Kear, G., Wood, R. J. K., Stokes, K. R., and Walsh, F. C. 2005. *Corr. Sci.* **47**, pp. 3336
[26] Musabikha, S., Utama, I. K. A. P., Mukhtasor, Wharton, J. A., and Wood, R. J. K. 2020. *Analyt. Let.* **5**, pp. 1
[27] Oldfield, J.. W, and Masters, G. L. 1996. *Collation of Data Comparing Properties of Aluminium Bronze with Cast Stainless Steels* (Hemel Hempstead UK: Copper Development Association)
[28] Walsh, F. C., Kear, G., Nahlé, A. H., Wharton, J. A., and Arenas, L. F. 2017. *Corr. Sci.* **123**, pp. 1

[29] Ault, J. 1995. *NACE International* (Houston) p. 1.
[30] Fraenkel, P. L. 2007. *Proc. Inst. Mech. Eng. Part A: J. Power & Energy*, **221**, pp.159
[31] Richardson, I. 2016. *Guide to Nickel Aluminium Bronze for Engineers* (Hemel Hempstead UK: Copper Development Association).
[32] Sandoval-Jabalera, R., Arias-del Campo, E., Chacon-Nava, J. G., Malo-Tamayo J M and Martinez-Villafane A 2006 *Portugal. Electroch. Acta,* **24**, pp. 393.
[33] AZO Material. 2013. *AISI 1045 Carbon Steel (UNS G10450)* (Manchester: AZoNetwork), p. 1
[34] Jadidi, O., Hong, T. S., Firouzi, F., Yusuff, R. M., and Zulkifli, N. 2008. *J. Achiev. Mater. & Manuf. Eng.* **31**, pp. 762.

IOP Conf. Series: Materials Science and Engineering 1158 (2021) 012002 doi:10.1088/1757-899X/1158/1/012002

Integrated Information System for Customer Care

D Riananingrum[1*], R R S Hari[2], F Nursaori[3], WA Astuti[4]

[1,2] Departemen Ilmu Komunikasi, Universitas Komputer Indonesia, Indonesia
[3,4] Departemen Akuntansi, Universitas Komputer Indonesia, Indonesia

Email: *devyriananingrum@mahasiswa.unikom.ac.id

Abstract. The application of customer Facing and Ecosystem Facing is implied as computerization of customer data. The goal is to understand customers' needs and expectations to establish good relationships with customers. This study aims to determine Customer Facing and Ecosystem Facing services in the digital transformation business. The research method used a descriptive research method with a qualitative approach. Customer Facing and Ecosystem Facing is a technology that can increase the production process for business. Salespeople are faster to create and send offers to clients because it provides CRM information.

1. Introduction

Technological developments allow automation in almost all fields. Industrial Revolution 4.0 is also the latest technology trend with a significant impact due to the growing development of artificial intelligence, electronic commerce, big data, financial technology, and the sharing economy, aiming to make human life and the environment better.

Technology cannot be separated from problems, because in essence, technology is born and developed to solve problems faced by humans. Therefore, technology has a role in expanding and enlarging human potential to meet practical needs [1].

The use of technology in companies should be balanced with digital literacy so that the benefits of technology itself can be absorbed optimally. Companies that have implemented or adopted technology will find it easier to carry out their company operations provided. Using technology, users must understand how to use it so that the tool or service's benefits can be maximally felt [2]. It shows that innovation in information technology and information systems has changed. Gartner defines digitization as "the use of digital technology to transform business models and provide new revenue and value-generating opportunities; it is the process of moving to a digital business" [3]. It means the use of digital technology must contribute to company performance [4].

Customer Relationship Management (CRM) is an organizational strategy that is customer-focused and customer-driven. It is an organization that concentrates on assessing customer requirements for products and services, then providing responsive and high-quality services [5]. CRM is not a process or technology; it is a customer-centered way of thinking and acting. CRM builds sustainable long-term customer relationships that create value for the company and the same customers [6].

Dealing with customers refers to the way customers experience or perceive business service features. CRM is a crucial component of customer-facing solutions designed to satisfy user experience through all customer contact points. Many customer-facing processes and technologies are significant business investment components with a strong influence on revenue generation [7].

CRM applications deal with customers, sales representatives, field, service, and customer interaction centers to interact directly with customers. These applications include customer service and support, sales force automation, marketing, and campaign management. One service data ecosystem uses areas and natural energy sources and spatial planning [8].

This study can also serve as an extension of previous research. The gaps found are related to the following: work stress caused by changes in the work system itself, intended to make the system more accessible and effective (1). However, if the company does not implement the strategy correctly, employees will experience work stress. which resulted in a hitch. (2). Radical changes can affect employees to become stressed in their work environment (3).

The application of customer Facing and Ecosystem Facing in fulfilling the needs of Digital Transformation examines the importance of strengthening good relationships between customers and employees to create expected work efficiency and service awareness using the Customer Facing Ecosystem Facing system. The research shows that due to the recent transformation of the digital era, this study explores the main challenges of digital change in good customer Facing and Ecosystem Facing services that will affect enormous consumer loyalty.

2. Method

Research methods can be interpreted as a scientific way to obtain valid data to find, develop, and be used to understand, solve, and anticipate problems in education [9]. The method used a descriptive research method with a qualitative approach. [10]. The qualitative research method is a research method based on the philosophy of postpositivism, which was used to examine a natural object's condition (as opposed to an experiment).

3. Results and Discussion

By using the customer management system, salespeople are faster to create and send offers to clients, which allows a quick follow-up without waiting for someone else to provide the information needed for the bid and make transactions in CRM. The indicators faced by customers are contact centers, sales force automation, and field service. It is called customer-looking because sales representatives, field service representatives, and contact centers are indicators that interact directly [11].

1. **Contact Center**
   Contact Centers are a network of contact centers maintained by network management servers that allow contact centers (which can be run independently of one another) to register via the internet or an intranet. Applications that use the telephone to support the CRM process include marketing, sales, and services. These applications include telemarketing and telemarketing [12]. Through customer relationship management applications, it is possible to manage their existing customer base and use these contacts to build an even larger group of potential clients. With CRM, customer information is in one centralized system. Selling to existing clients is much more effective than selling to new clients.

2. **Sales Force Automation (SFA)**
   Applications in SFA help companies carry out sales force activities, manage prospects, and customers in the sales process. A salesforce automation system that integrates an intelligent, computerized sales force supports multiple sales process phases [13]. Through a cloud-based CRM system, sales professionals will be able to complete their work wherever they are as long as there are Wi-Fi and a device to access the internet. This software helps the business relationship between the company and the client and the salesperson's daily tasks.
3. **Field Service Automation (FSA)**
   The FSA application supports the service efforts of field service departments and service managers. Effective service logistics can reduce costs and increase service value by increasing

customer satisfaction and loyalty. This application manages services from customer requests, orders, contracts, schedules, and calls. [14]. Automation is a technique or system that runs or controls fully automatic tools with electronic means to reduce human labor use.

3.1. Facing the Ecosystem

Many companies started to transform into collaborative information technology ecosystems for survival - especially the traditional offline manufacturing and service industries. Business models are constantly changing and evolving. The business model from World War II to 1980 was a combination of operations across industries. Since the 1980s, companies have been experts focusing on only one sector. In the future, as the collaborative ecosystem grows, industry boundaries are blurred. Physical products can be connected to the online world through the Internet of Things. Traditional manufacturing and service industries can now access their product data. Apart from that, customers can also purchase this product through an access method that was never thought of before. With access to large and diverse amounts of data, existing companies can create value-added products and forge partnerships across multiple industries. Therefore, a new ecosystem model was developed for incumbents who previously only ran their business offline.

Digital ecosystems are gaining momentum in the business landscape, ushering in a new value chain evolution era. The challenges are significant, but the companies who face them and seize the power of platforms will be rewarded with the keys to future markets. All over the world, digital ecosystems are growing and flourishing, fundamentally changing what it means to add value. The ascendancy of native platform companies demonstrates this perfectly: by shaping new business ecosystems around themselves, they create transformative benefits for both their partners and the markets they serve.

Ecosystem nowadays belongs to the digital world with the customer-focused model, where users can enjoy an end-to-end experience. The ecosystem will consist of various participants providing multi-industry solutions that are digitally accessible. This ecosystem relationship will enable the company to meet customer expectations better. The internet, advanced data analysis functions, and the maturity of Artificial Intelligence (AI) allow consumers to find the best solution in just a few milliseconds. These dynamics are taking place in the high tech, media, and telecommunications industries and are supported by tech giants who have built platforms that can destroy entire ecosystems. To meet the growing trend of customer expectations, the company is expanding its product and service range in an unprecedented way. They also form alliances with other companies (even with their competitors) to create complementary products and services.

Customer-facing ecosystems are interconnected corporate networks within the company. Accenture identifies five different options for organizations to develop and launch ecosystems [15,16]:

1. The live moment of the orchestra
   With this type of ecosystem, banks can organize the ecosystem around specific life moments, offering partners access to their customer base in exchange for a fee.
2. Market regulators
   This type of ecosystem allows the bank to operate as a market orchestrator through white labeling or co-branding. Market authors sell non-financial products to customers.
3. Participants of the third-party ecosystem
   Third-party ecosystems are where banks can join third-party platforms to offer banking products to third-party customers.
4. Open the company platform
   An open banking platform allows bank partners to incorporate specific products, data, or processes into their value proposition. With this ecosystem, partners can take advantage of open Application Programming Interfaces (APIs).
5. Reference platform
   This ecosystem directs customers to other providers, such as large banks that send small businesses to designated online financial platforms. However, this ecosystem is not hosted on third-party platforms.

4. Conclusion

The concept of ecosystem services was initially developed to illustrate the benefits that natural ecosystems generate for society and raise awareness for biodiversity and ecosystem conservation. Customer Relationship Management (CRM) is an integrated information system used to schedule, control, and plan pre-sales and post-sales activities in an organization/company. CRM itself is closely related to Customer-Facing and Ecosystem Facing services to meet the needs of digital transformation. Thus, this research contributes to comprehensive results that connect the customer-face and ecosystem faced to fulfill digital transformation if carried out with good strategy and implementation in the work environment. However, good service to customers affects the company's image. The customer and the ecosystem's face is a significant tool/relationship to convey an impression for the tool/system.

Acknowledgement

The author would like to thank Prof. Dr. Ir. H. Eddy Soeryanto Soegoto, MT as a lecturer in Entrepreneurship in Accounting and Communication at the Universitas Komputer Indonesia, who provided insight and advice for this research.

References

[1] Anatan, L., & Ellitan, L. 2009. *Manajemen Inovasi (Transformasi. Menuju Organisasi Kelas Dunia).* Bandung: CV. Alfabeta.

[2] Legner, C. et al. 2017. *Digitalization: opportunity and challenge for the business and information systems engineering community. Business & information systems engineering*, ***59***(4), pp. 301-308.

[3] Gartner Digitalization Retrieved 18.11.2018 from https://www.gartner.com/itglossary/digitalization/

[4] Riedl, R., Benlian, A., Hess, T., Stelzer, D., Sikora, H. 2017. *On the relationship between information management and digitalization. Business & Information Systems Engineering,* ***59***(6), pp. 475-482.

[5] Schwertner, K. 2017. Digital transformation of business. *Trakia Journal of Sciences*, ***15***(1), pp. 388-393.

[6] Rababah, K., Mohd, H., & Ibrahim, H. 2011. Customer relationship management (CRM) processes from theory to practice: The pre-implementation plan of CRM system. *International Journal of e-Education, e-Business, e-Management and e-Learning,* ***1***(1), pp. 22-27.

[7] Riqqi, A., Hendaryanto, H., Safitri, S., Mashita, N., Sulistyawati, E., Norvyani, D. A., & Afriyanie, D. 2019. Pemetaan jasa ekosistem. In *Seminar Nasional Geomatika* ***3***, pp. 237-246.

[8] Sugiyono. 2016. Memahami Penelitian Kualitatif. Bandung: Alfabeta.

[9] Johnson, J. S., Matthes, J. M., & Friend, S. B. 2019. Interfacing and customer-facing: Sales and marketing selling centers. *Industrial Marketing Management*, *77*, pp. 41-56.

[10] Delaney, P. J. 2005. U.S. Patent No. 6,937,715. Washington, DC: U.S. Patent and Trademark Office.

[11] Johnson, J. D., Lundberg, D. R., & Krebsbach, M. P. 2009. U.S. Patent No. 7,516,088. Washington, DC: U.S. Patent and Trademark Office.

[12] Cheung, C. F., Chan, Y. L., Kwok, S. K., Lee, W. B., & Wang, W. M. 2006. A knowledge-based service automation system for service logistics. *Journal of Manufacturing Technology Management.*

[13] https://www.fintechmagazine.com/banking/new-digital-age-banking-ecosystems

[14] https://www.techopedia.com/definition/23325/customer-facing

[15] Introduction to Information Systems, 5th Edition Int'l Student Version

IOP Conf. Series: Materials Science and Engineering 1158 (2021) 012002 doi:10.1088/1757-899X/1158/1/012002

[16] http://blogs.msdn.com/b/domc/archive/2008/04/28/customer-facing-application- design-a-paradigm-shift.aspx

IOP Conf. Series: Materials Science and Engineering 1158 (2021) 012003 doi:10.1088/1757-899X/1158/1/012003

Comparative Performance of Induction Motor Speed Controller System with Flux Weakening Control

Regina Chelinia Erianda Putri[1], Feri Yusviar[2]
[1,2]Electrical Engineering Department, Faculty of Electrical Engineering, Universitas Indonesia, Indonesia

Email: regina.chelinia@ui.ac.id(1), yusivar@ee.ui.ac.id(2)

Abstract. In general induction motor not work optimally because cannot reach the maximum speed of the rating. In this paper, simulation is used with several methods, named Maximum Torque Per Ampere (MTPA) and Field-weakening. Simulation of the methods used the function on MATLAB C-Mex and implemented on block of Simulink. The Field-weakening is control speed of the induction motor by decreased flux. The goal is to produce motor speed above the rating. The method of field-weakening control with addition of $|i_q|$ used as reference i_d^*, which will be used to limit i_q^*. The result of the study is a comparison between the two methods to verify reliability. Simulations results showed that induction motor speed controller using field-weakening control with addition $|i_q|$ better performance.

1. Introduction

The control method for optimization of characteristics on the performance of the induction motor grows fast. Various used induction motors cause there are various methods in matching the desired performance and specifications. One method is the Field-Weakening Method which is often used to increase motor speed above rating. This Field-Weakening method purposed to control i_q^* by limiting it i_d^* where there is $|i_q|$ input [1]. There is MTPA method (Maximum Torque Per Ampere) where in this method purposed to ensure that the current used is the minimum i_d current for development on motor torque, with total flux stator controlled by MTPA criteria [2].

There is some research on Field-Weakening and MTPA methods. The Field-Weakening Method already has several modifications such as the *"Predictive Control Based Field-Weakening Strategy for Traction EV Used Induction"* [3] which used Predictive Control (MPC) based control where the MPC method is adopted to replace the PI regulator in the voltage loop to calculate the current reference that produces flux. Second, the title is *"Overmodulation and Field Weakening in Direct Torque Control of Induction Motor Drives"* [4] which used the Overmodulation method where this method only modifies torque errors and flux before fed to the look-up table. The third, the title is *"High Performance Induction Motor Drive in Field Weakening Region"* [5] which in the calculation of field-weakening method used iron losses. MTPA method has several various developments such as the title *"Maximum Torque-per-Amp Tracking Control of Saturated Induction Motors"* [7] which proposed to improve the torque-flux tracking performance and controller robustness. The next reference is titled *"A Maximum Torque per*

Ampere Control Strategy Drives" [8] which is proposed simply in structure and has the straightforward goal of minimizing the stator current amplitude for a given load torque.

This paper will compare the performance of induction motor system control method with several methods namely, MTPA (Maximum Torque Per Ampere), Flux-Weakening with the addition of $|i_q|$ and the combination of MTPA and Flux-Weakening Control.

2. Implementation of Controls with Matlab

Model of the induction motor system used in the comparison is the same for each method. The difference is found in the speed control adjusted to each method. It is a Rotor Flux Oriented Control (RFOC) of an induction motor (a three-phase, four-poles, 750 W, squirrel-cage induction motor)

2.1. *Induction Motor Model*

On the controller, the scheme requires a drive induction motor model. Induction motor models used in all methods used a synchronous frame of reference, with the following equation: [10]

$$\frac{d}{dt} i_{\alpha s} = \frac{R_s L_r i_{\alpha s} - N_p \omega_r L_m^2 i_{\beta s} - R_r L_m i_{\alpha r} - N_p \omega_r L_r L_m i_{\beta r} - L_r v_{\alpha s}}{L_m^2 - L_r L_s} \tag{1}$$

$$\frac{d}{dt} i_{\beta s} = \frac{N_p \omega_r L_m^2 i_{\alpha s} - R_s L_r i_{\beta s} - N_p \omega_r L_r L_m i_{\alpha r} - R_r L_m i_{\beta r} - L_r v_{\beta s}}{L_m^2 - L_r L_s} \tag{2}$$

$$\frac{d}{dt} i_{\alpha r} = -\frac{R_s L_m i_{\alpha s} - N_p \omega_r L_m L_s i_{\beta s} - R_r L_s i_{\alpha r} - N_p \omega_r L_r L_s i_{\beta r} - L_m v_{\alpha s}}{L_m^2 - L_r L_s} \tag{3}$$

$$\frac{d}{dt} i_{\beta r} = -\frac{N_p \omega_r L_m L_s i_{\alpha s} + R_s L_m i_{\beta s} + N_p \omega_r L_r L_s i_{\alpha r} - R_r L_s i_{\beta r} - L_m v_{\alpha s}}{L_m^2 - L_r L_s} \tag{4}$$

Where Ls, Lr, and Lm show each phase induction of stators, rotors. Whereas σ is defined as $1 - \frac{Lm^2}{Ls\,Lr}$. φ_{ds} and φ_{qs} the components d and q axes on the flux stator parallel along the rotor flux axis.

2.2. *MTPA (Maximum Torque Per Ampere)*

In the MTPA method, the system is modified in such a way that the MTPA controller is designed along with a feedback-linearizing controller. The blok diagram of MTPA control is as shown in Figure 1.

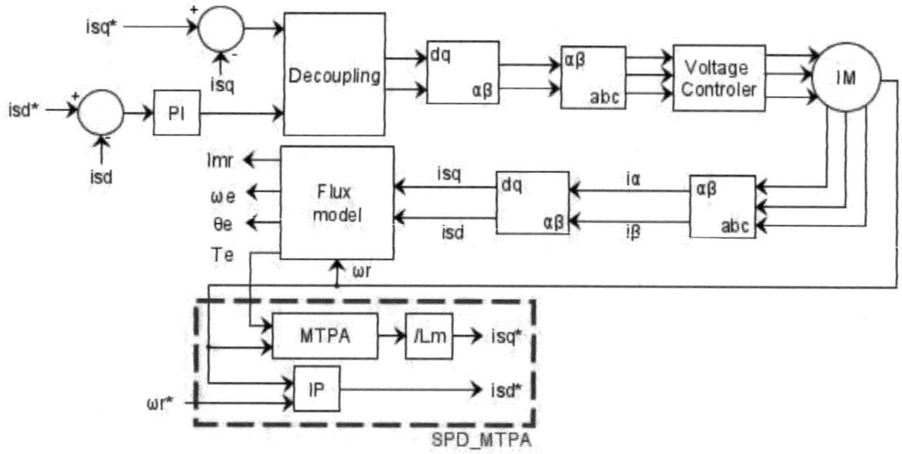

Figure 1. Blok Diagram of MTPA (Maximum Torque Per Ampere)

In the process of identified MTPA conditions for specific motor torque, the current voltage is determined as a function of the d and q axes, so that the stator current components used are i_d and i_q. Therefore, the proposed control used T_e in its control so on the MTPA method, Te shows the electromagnetic torque of the equation given by (5),

$$T_e = p(1-\sigma)L_s\, i_{sq}^*\, imr \tag{5}$$

MTPA operated conditions on each of the d and q axis flux components within the synchronous reference framework can be provided by (6) and (7)

$$\varphi_{ds}|_{MTPA} = \frac{L_s}{L_m}\sqrt{\frac{2T_eL_r}{P}} \tag{6}$$

$$\varphi_{qs}|_{MTPA} = \frac{\sigma L_s}{L_m}\sqrt{\frac{2T_eL_r}{P}} \tag{7}$$

The values of the d and q flux axis components within the synchronous reference framework can be given by (8),

$$|\varphi_s|_{MTPA} = \frac{L_s}{L_m}\sqrt{\frac{2L_r\,(1+\sigma^2)T_e}{P}} \tag{8}$$

2.3. *Flux-Weakening added by* $|i_q|$

In the conventional field, the induction motor-oriented control flux rotor is assumed to be constant so that torque can be controlled by controlling the current of the q axis. Weakens the flux to control the motor so that it can reach very high speeds, although it has a reduced risk of torque. This method adds i_q current to the reference on the d-axis so that the right flux and torque can be generated automatically. Block diagram of the field-weakening method with the addition of $|i_q|$ shown by Figure 2.

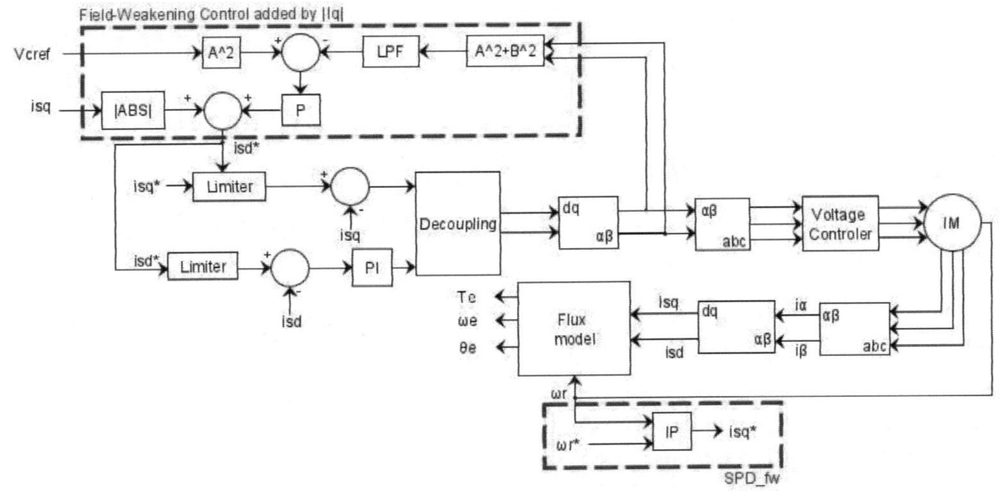

Figure 2. Blok Diagram of Flux-Weakening added by |Iq|

The operation is performed by limiting the current on the d (i_{sd}) and q (i_{sq}) axes which will limit the reference current on the d (i_{sd}^*) and q (i_{sq}^*) axes. Application of optimization restrictions is carried out Field weakening techniques by determining the reference d-axis current i_{sd}^* which is limited to the rated value with the equation (9)

$$|i_{s_{max}}| \geq \sqrt{i_{sd}^2 + i_{sq}^2} \quad (9)$$

The voltage applied to the motor is limited, so the V_{dc} used is limited to $m.V_{dc}/2$ where m is index modulation this method used a filter to generate reference current on the d-axis (i_{sd}^*), which is expressed by equation (10)

$$i_{sd}^{e*} = K_{vp}(V_c^{*2} - v_s^{filter} + K_{abs}|i_{qs}^e|) \quad (10)$$

Where K_{vp} represent proportional amplifier, v_c^* is the set V_{dc} voltage, v_s^{filter} is the filtered signal, while $K_{abs}$is the absolute q axis current gain. Besides i_d^* is fed to the d-axis current controller, it is also used to determine i_{qmax} performing the current limiter as in (9).

2.4. Combined MTPA plus Flux-Weakening Control

In this method, the motor is controlled by combining the two previous methods. The block diagram of the system is shown in Figure 3.

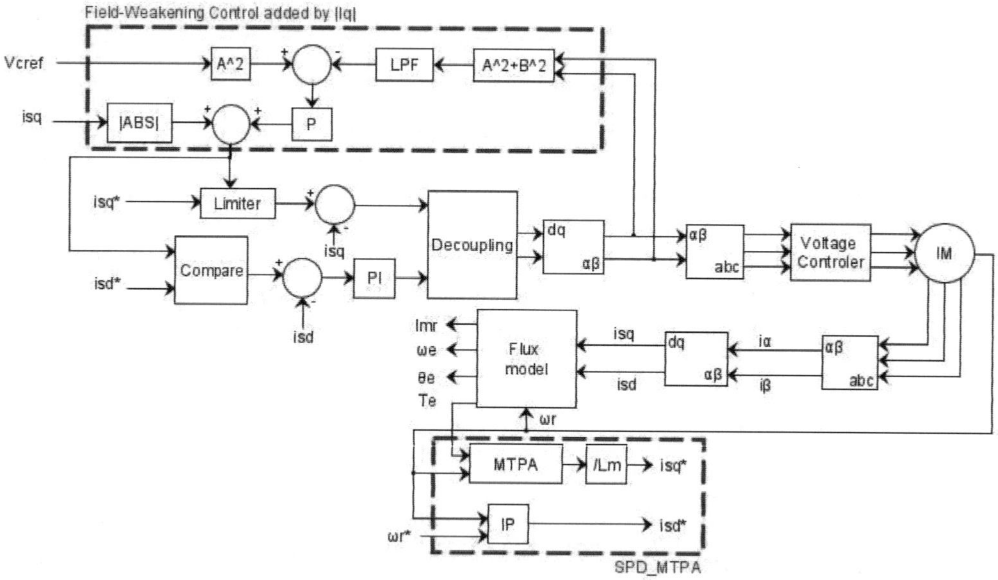

Figure 3. Block Diagram of MTPA with Flux-Weakening Control

In this method, the MTPA controller is used to provide one of the references in the i_{sd}^*, where the i_{sd}^* used is the one that has the lowest value followed by the PI gain. Field-weakening control is used as reference for i_{sq}^*.

3. Experimental Results Of Control Methods

3.1. Experimental Settings

The proposed methods are simulated using MATLAB SIMULINK. Applied the algorithms for this system to an experimental system implemented by the embedded 'C' code. Parameters used are listed in Table 1.

Table 1. Parameters of IM

Parameter	Symbol	Value
Number of pole pairs	P	2
Stator Resistance	R_s	2.76

IOP Conf. Series: Materials Science and Engineering 1158 (2021) 012003 doi:10.1088/1757-899X/1158/1/012003

Parameter	Symbol	Value
Rotor Resistance	R_r	2.9
Stator Inductance	L_s	0.2349
Rotor Inductance	L_r	0.2349
Mutual Inductance	L_m	0.2279

The value of the gain on the speed controller of each method is equalized, used $K_i = 0.9$ and $K_p = 0.2$ while for field-weakening is given a gain of K_p=0.0009. Comparative simulations are conducted used 2 values ω_r^* =100 (955 rpm) and ω_r^* =300 (2865 rpm).

3.2. Method Comparison Results

The results of the MTPA, Field-weakening and combination of MTPA and Field-weakening method experiments are as shown sequentially in Figures 4-6.

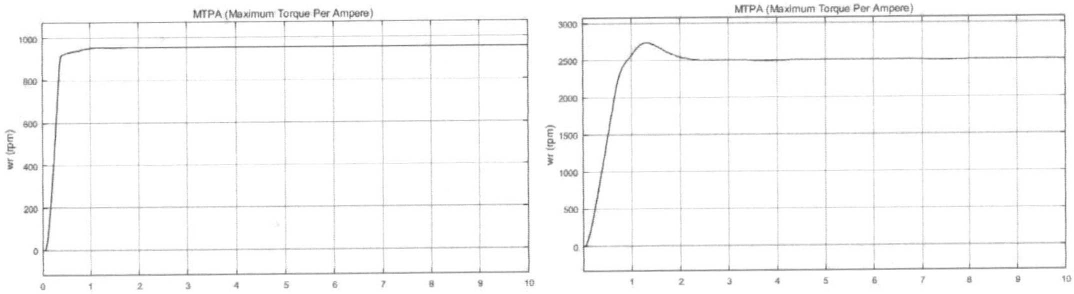

(a) (b)

Figure 4. (a) MTPA in ω_r^* = 954 rpm , (b) MTPA in ω_r^* =2865 rpm

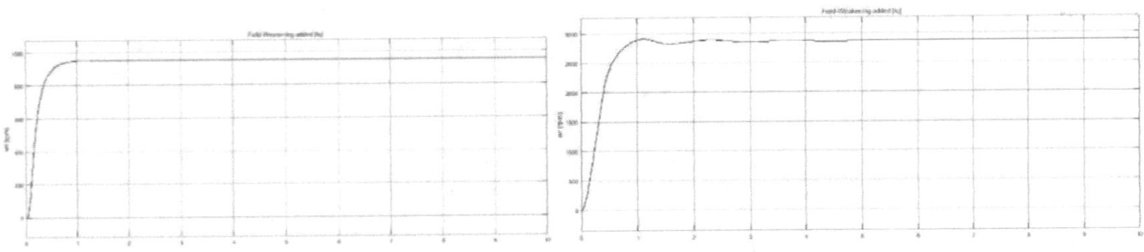

(a) (b)

Figure 5. (a) Field-weakening in ω_r^* =954 rpm, (b) Field-weakening in ω_r^* =2865 rpm

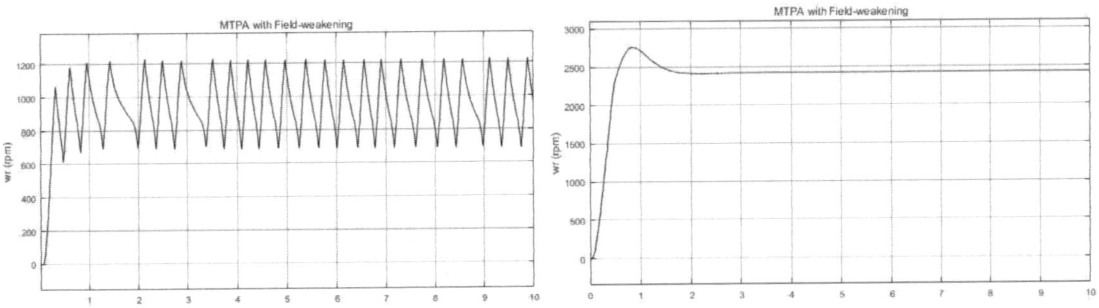

(a) (b)

Figure 6. (a) MTPA with Field-weakening in ω_r^* =954 rpm, (b) MTPA with Field-weakening in a ω_r^* =2865 rpm

Comparison of three method results as shown in Figure 4(a), Figure 5(a), Figure 6(a) at ω_r^* =955 rpm, with a risetime of 233,809 ms. The MTPA method performs best among others. This is because the MTPA method reaches a value of 954 rpm with the minimum oscillation and the fastest settling-time in reaching its steady-state. The Field-weakening method with T=10 still produced good graphics but not as fast as the MTPA method, which reaches 954 rpm with a risetime of 348.8 ms. While in the combination of MTPA method with Field-weakening with T=10 the system still continues to oscillate and does not reach steady state.

Furthermore, as shown Figure 4(b), Figure 5(b), Figure 6(b) at ω_r^* =2865 rpm MTPA method seen in the system oscillates but can not reach the desired speed of about 2492 rpm with a risetime of 561,725 ms. The Field-weakening method with T = 10 has the best results, the system looks oscillating but already in the steady-state range. Field-weakening method can reach its steady state at 2864 rpm with the required risetime time of 497,851ms. While the MTPA combination method with Field-weakening has different response compared to the condition ω_r^* =955 rpm, at ω_r^* =2865 rpm, in the system there is a slight oscillation and achieved a steady-state condition but not reached the desired speed. In the combination method, the system reached 2421 rpm with a risetime of 330,989 ms.

Comparison of three method results confirmed that Field-weakening method with added $|i_q|$ has the best performance for high speed at ω_r^* =2865rpm.

4. Conclusion

In all three methods proposed to be simulated on induction motor, it was obtained that the response of field-weakening method with the addition of $|i_q|$ have the best performance among the two other methods. The response of MTPA combination method with field-weakening does not produce better condition compared to field-weakening method with the addition of $|i_q|$. The addition of higher voltage can be done in the field-weakening method. Addition of $|i_q|$ can provide stimulation to the motor to increase the resulting flux thus affected the flux and torque of the induction motor

Acknowledgement

The authors would like to thank all who always supported and all that involved in this research so that the Author can complete this research.

References

[1] Yusivar, Feri, et al. 2001. "| Iq| added flux weakening strategy for the rotor flux oriented control of a sinusoidal PWM VSI-fed induction motor." *IECON'01. 27th Annual Conference of the IEEE Industrial Electronics Society (Cat. No. 37243)*. **2**. IEEE

[2] Naganathan, Pratibha, and Srirama, S. 2018. Maximum Torque Per Ampere Based Direct Torque Control Scheme of IM Drive for Electrical Vehicle Applications.*2018 IEEE 18th International Power Electronics and Motion Control Conference (PEMC)*. IEEE

[3] Su, J., Gao, R., and Husain, I. 2017. Model predictive control based field-weakening strategy for traction EV used induction motor. *IEEE Transactions on Industry Applications*, **54**(3), pp.2295-2305.

[4] Ahmad, M. Z., and Idris, N. R. N. 2006. Overmodulation and field weakening in direct torque control of induction motor drives. *In 2006 IEEE International Power and Energy Conference*, pp. 465-469. IEEE.

[5] Aswathy, M. S., and Beevi, M. W. 2015. High performance induction motor drive in field weakening region. *In 2015 International Conference on Control Communication & Computing India (ICCC),* pp. 242-247. IEEE.

[6] Kouns, H., Lai, J. S., and Konrad, C. E. 2004. Analysis of a traction induction motor drive operating under maximum efficiency and maximum torque per ampere conditions. *In Nineteenth Annual IEEE Applied Power Electronics Conference and Exposition, 2004. APEC'04*. **1**, pp. 545-551. IEEE.

[7] Peresada, S., Kovbasa, S., Dymko, S., and Bozhko, S. 2017. Maximum torque-per-amp tracking control of saturated induction motors. *In 2017 International Conference on Modern Electrical and Energy Systems (MEES)*, pp. 72-75. IEEE.

[8] Wasynczuk, O., Sudhoff, S. D., Corzine, K. A., Tichenor, J. L., Krause, P. C., Hansen, I. G., & Taylor, L. M. 1998. A maximum torque per ampere control strategy for induction motor drives. *IEEE Transactions on Energy Conversion*, **13**(2), pp.163-169.

[9] Kim, S. H., and Sul, S. K. 1995. Maximum torque control of an induction machine in the field weakening region. *IEEE Transactions on Industry Applications*, **31**(4), pp.787-794.

[10] Yusivar, F., and Wakao, S. 2001. Minimum requirements of motor vector control modeling and simulation utilizing C MEX S-function in MATLAB/SIMULINK. In 4th IEEE International Conference on Power Electronics and Drive Systems. *IEEE PEDS 2001-Indonesia. Proceedings (Cat. No. 01TH8594),* **1**, pp. 315-321. IEEE.

[11] Hamada, F., Husnayain, F., and Yusivar, F. 2019. Speed Sensorless Vector Control of Parallel Connected Induction Motor with Anti-windup Integral-Proportional Speed Controller. *In 2019 IEEE 2nd International Conference on Power and Energy Applications (ICPEA)*, pp. 89-93. IEEE.

[12] Kono, Y., Fushimi, T., and Matsuse, K. 2000. Speed sensorless vector control of parallel connected induction motors. In Proceedings IPEMC 2000. *Third International Power Electronics and Motion Control Conference (IEEE Cat. No. 00EX435)*, **1**, pp. 278-283. IEEE.

IOP Conf. Series: Materials Science and Engineering 1158 (2021) 012004 doi:10.1088/1757-899X/1158/1/012004

A Modeling of Lead-Acid Battery for Electric Vehicles and Simulation Utilizing C MEX S-Function in Matlab/Simulink

A D Kheirisko[1], F Yusivar[1]
[1]Electrical Engineering Master Program, Faculty of Electrical Engineering, Universitas Indonesia, Indonesia

e-mail : allbowaghis.digandra@ui.ac.id

Abstract. Nowadays, the need for better performance of batteries as an energy source is increasing following the growth of electrical vehicles (EVs). Therefore, modeling the battery system is important to analyze how the battery is able to perform better so that further increases the EV's performance. This study is addressed to analyze and simulate circuit modeling of lead-acid battery into a pretty complex RLC circuit, voltage source of battery itself, and a connected load with provided conditions for obtaining varying parameters which are used by the battery with a permanent initial state of charge (SoC). This circuit is simulated using C/C++ programming where it is implemented to some connected C MEX S-Function blocks in Simulink. The result of the study is proved that series circuits increase their voltage and parallel circuits increase their ampere. For satisfying the EV's specs such as V_{out} = 400 V and the capacity = 100 Ah, arranging batteries in 400 series since n = 15 produce around 13 V.

1. Introduction

Battery is a very essential component to many fields such as electrical tools, mechanical tools, and so on. And, one example of battery usage in electrical tools is the battery as the main power source to operate Electric Vehicles (EVs). The battery's usage as the main power source in EVs directly states the importance and vital influence of the battery's performance [1]. One battery type used in EVs is the lead-acid battery. The lead-Acid battery is accessible and has the highest maximum voltage value in its category [2]. Lead-Acid batteries are typically used in the automotive industry, where they undergo periodic deep discharge and charge, and in stationary applications, where they remain in charged state most of their life [3]. Since EVs require decent battery performance, it is important to model a battery that is suitable for EVs characteristics. The value chain of EVs batteries consists of seven steps: component production (including raw materials); cell production; module production; assembly of modules into the battery (including an electronic control tuning and a cooling system); integration of the battery pack into the vehicles; used during the life of the vehicle; and reuse and recycling [4]. Before running the processes mentioned above, it is important to simulate the battery model before applied in physical EVs. The simulation can be run using various methods and software. One of them is Matlab Simulink using C-Mex S-Function, which is applied using the programming language. The program adopted the derivative equation of the circuit from [5].

A recent study has underlined the importance of a standard battery module to achieve massive electrification of off-road vehicles [6]. According to the study, it is possible to use a standard module of 4 series-connected cells to build up the battery of many types of off-road vehicles [7]. The standard module has a 12.8 V nominal voltage. Its implementation with three capacity values: 30, 60, and 100 Ah is reported [8]. The large numbers of cells are connected to construct a battery pack to satisfy the voltage and capacity requirement of an electric vehicle [9]. Cell failure and imbalance are critical problems in battery storage systems, especially in series-connected battery strings [10]. So, we need a simulation for knowing and observing batteries' behaviour, particularly in series-parallel connections. In this study, the simulation will be observed in Matlab Simulink using C-Mex S-Function. It is a novel process since the simulation mostly will take part in coding using C language, so we need more knowledge about C and information about the battery, series-parallel connections, and others.

The general battery used is MSE-100Ah with a nominal voltage of 6 V and a capacity value of 100Ah. Once the required voltage, current and capacity are defined for the target application, it is easy to calculate the number of necessary cells. The battery voltage is set by the number of series-connected cells, whereas the battery capacity and the maximum battery current are given by the number of parallel-connected cells. In [5], the equivalent circuit below is the so-called Randles equivalent circuit with an additional inductance.

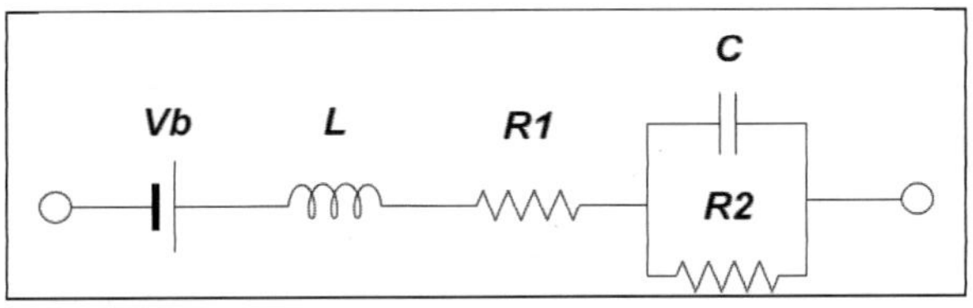

Figure 1. Equivalent Circuit of Lead-Acid Battery[5]

Usually, the equivalent circuit for batteries uses only the Randles equivalent circuit. In this case, the inductance of the storage battery can be said as the form of the energizing part of the storage battery. The inductance occurs due to the wiring straps between the cell element and battery terminal/pole and between one cell element and the other cell element inside the battery. This paper also used an electric circuit parameter at each state of charge in [5].

Table 1. Electric Circuit Parameter at Each State of Charge

State of Charge	R_1 (mΩ)	R_2 (mΩ)	C (F)	L(μH)
100%	1.92	0.9	12	0.75
80%	2.06	0.81	13.1	0.81
60%	2.2	0.69	13.6	0.78
40%	2.28	0.61	12.5	0.68
20%	2.72	0.67	11.1	0.79
0%	3.06	1.06	4.25	0.85

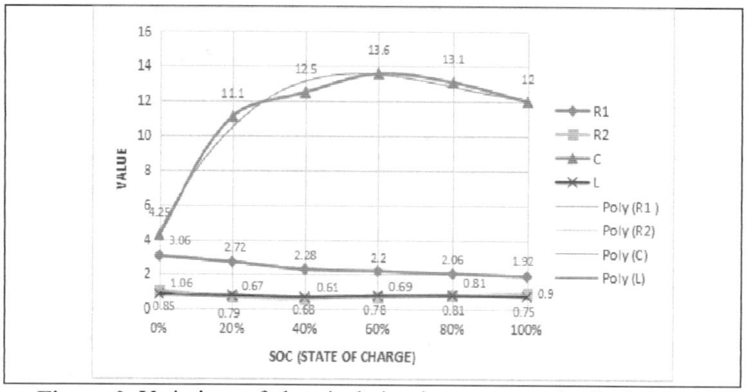

Figure 2. Variations of electrical circuit parameters in discharging.

The processes that will be explained are to derive a circuit to the state-space equation that can be transformed to the discrete domain and then implemented to C-Mex S-Function.

2. Method

2.1 Mathematical Model of Lead-Acid Battery

According to the equivalent circuit in Figure 1, the equation to determine voltage value of the battery (V_{out}) is as below:

$$V_{out} = V_C + V_L + V_B + V_{R1} \quad (1)$$

Where V_c, V_L, V_B, and V_{R1} is voltage of capacitance component C, inductance component L, battery internal voltage, and resistance R_1 component respectively.

The value of the V_{out} can be determined by using state-space equation below:

$$\begin{bmatrix} \dot{V}_C \\ \dot{I}_L \end{bmatrix} = \begin{bmatrix} \frac{-1}{(C \times R_1)} & \frac{-1}{C} \\ 0 & 1 \end{bmatrix} \begin{bmatrix} V_C \\ I_L \end{bmatrix} \quad (2)$$

From the state-space equation above in mind, we can have equation to calculate V_c as below:

$$(V_c)_n = \Sigma_{n=0} \left[(V_c)_{(n-1)} + \frac{-1}{C} (I_{batt})_{(n-1)} + \frac{V_c}{R_2} \right] \quad (3)$$

Where,

$I_{out} = I_L = I_{batt}$

C = Capacitance value of component C

R_2 = Capacitance value of component R_2

R_1 = Capacitance value of component R_1

I_{out} is the output current of the battery, I_L is the current that pass-through inductance component L, and I_{batt} is current value inside the battery. The equation above will be written in C programming language in discrete structure using sampling period Δt which is equal to 10^{-3} seconds. The $(V_c)_n$ have to be multiplied to Δt.

To determine V_L, we use equation as below:

$$(V_L)_n = \Sigma_{n=0} \left[-L \left(I_{batt} - (I_{batt})_{(n-1)} \right) \right] \quad (4)$$

Where,
L = Capacitance value of component L. The $(V_L)_n$ have to be divided by sampling period time Δt.

To determine the value of V_{R1}, the equation below is used:

$$V_{R1} =- I_{out} \times R_1 \quad (5)$$

We also need to determine the State-of-Charge SoC value of the Lead Acid battery. To determine SoC_k we are using equation below:

$$SoC_k = \sum_{k=0} \left[SoC_{(k-1)} - \left(\frac{n}{Q}\right) \left(I_{batt}\right)_k \right] \quad (6)$$

Where,
n = battery cells
Q = battery capacity
The SoC_k has to multiply to Δt. From Figure 2, we did interpolation process to generate 3rd order polynomial equation for each parameter of R1, R2, C, and L. Below are the equation of each parameter:

$$R_1 =- 0.8796x^3 + 2.3016x^2 - 2.5728x + 3.0797 \quad (7)$$

$$R_2 =- 2.4306x^3 + 5.0387x^2 - 2.7677x + 1.0555 \quad (8)$$

$$C = 23.553x^3 - 56.468x^2 + 40.596x + 4.4329 \quad (9)$$

$$L =- 1.2037x3 + 2.0556x2 - 0.9582x + 0.8632 \quad (10)$$

2.2 Cmex S Function
The mathematical model that was previously discovered is able to be run in simulink using Cmex S-Function block. Cmex S-Function is a feature provided by simulink so it is possible to build a block in simulink using code created in MATLAB, C, C++, or Fortran. Other than MATLAB, it can be compiled using another compiler which one of them is MinGW Compiler. Compiling C, C++, or Fortran can be done using the mex command on the available command window. Cmex S-Function also uses callback methods on its structures that consist of mdlIntializeSizes, mdlIntiliazeSampleTimes, mdlOutputs, and mdlTerminate. MdlInitializeSizes specifies the sizes of various parameters in the SimStruct, such as the number of output ports for the block. MdlInitializeSampleTimes specifies the sample time(s) of the block. MdlOutputs calculates the output of the block. MdlTerminate performs any actions required at the termination of the simulation. If no actions are required, this function can be implemented as a stub. The Cmex routines will be shown in Figure 3 clearer.

To code the Cmex S-Function in C programming language, there are three provided structures, which are structure A, B, and C. Each of them has its characteristic: structure A accommodates calculation in discrete or continuous routines. Therefore, it does not mdlUpdate nor mdlDerivatives function since it does not have any state as well. The structure B accommodate calculation in discrete, and the most notable characteristic of structure B is it has any discrete states, and it has mdlUpdates function, which contains integral equation that will be used, which named solver algorithm since MATLAB is not able to process integral function system. In this structure, we must describe the time sampling of the process. This structure commonly figured for the controller program of a system. On the other hand, structure C accommodates calculation in continuous routines, and its main attribute is the structure have any continuous states, and it has mdlDerivatives function, which contains the differential equation, so that we don't need to write the solver or the integral equation, since, MATLAB will act as the solver. There are many kinds of the solver such as ODE45, ODE23, ODE Bogacki and so on. This structure is mainly used for the plant program of a system. These structures (A, B, and C) can be chosen depending on the equation form and the required output of the system. These explanations could be simplified in Figure 3 and Figure 4.

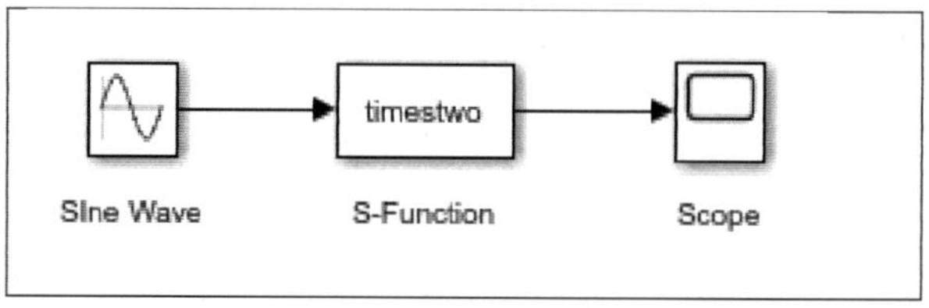

Figure 3. Block Diagram for The S-Function

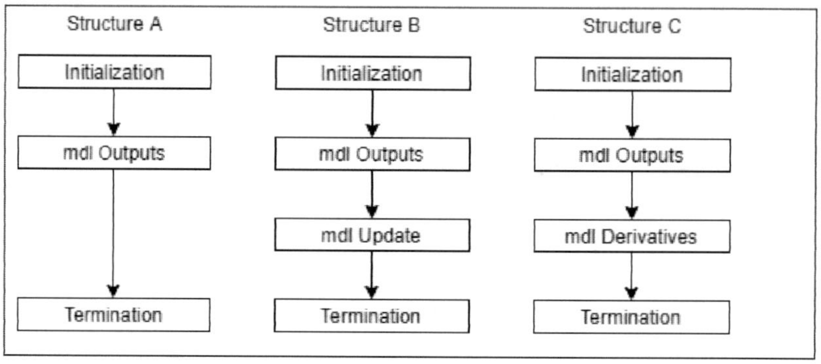

Figure 4. Cmex Routines

2.3. Implementation

The explanation from the previous section can be implemented to a simulation described in Figure 5.

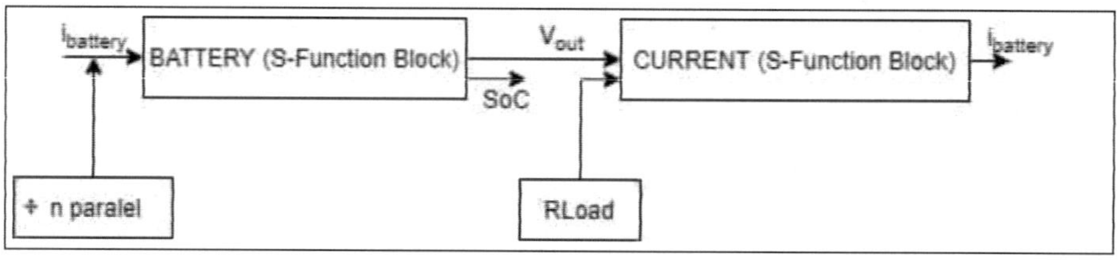

Figure 5. Parallel configuration s-function block

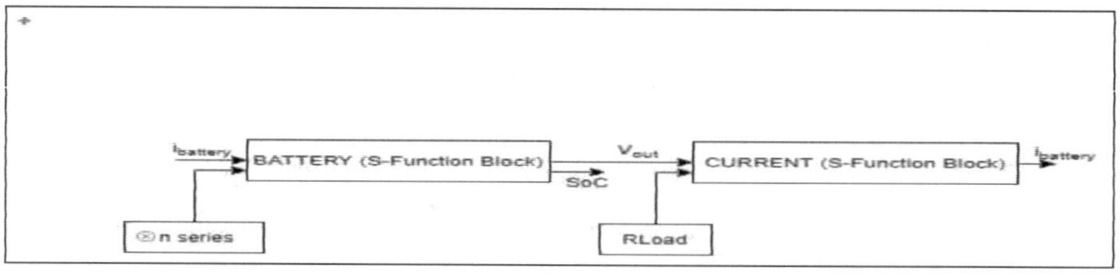

Figure 6. Series configuration s-function block

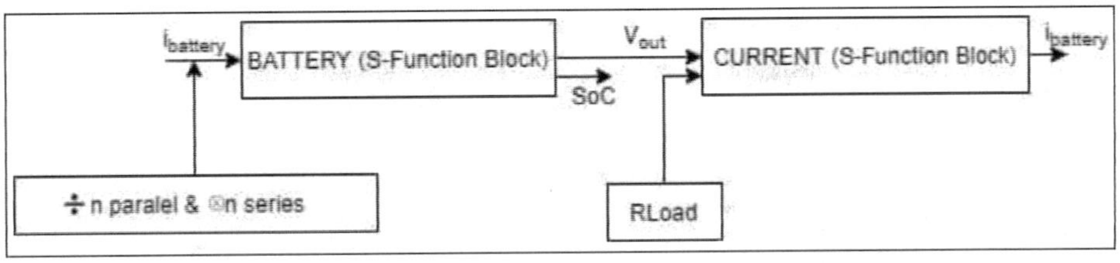

Figure 7. Series-Parallel configuration s-function block

The simulation process starts from 'BATTERY's-function block. In this s-function block, there exists a programmed process that runs using Cmex S-Function. In more detail, this Cmex S-function block has one input which is current feedback from a 'CURRENT' S-Function block and it has five output which are i_{act}; SoC; V_{out}; and R_1, R_2, C, and L to validate parameter value using 3rd order polynomial equation that had obtained. Then, Vout result is set as first input of the 'CURRENT' block, and a preset value of resistance is set as second input of the block. The output of the block is V_{out}, $I_{battery}$ value to monitor and validate calculation results of the block. The calculation and equation inside the 'CURRENT' block using Kirchoff Law no. 1 which is:

$$V = iR \quad (11)$$

From Figure 5 to 7, we can go further to determine batteries arrangement whether parallel, series, or series-parallel. It is helpful to have the best battery arrangement, so that it is compatible and suitable to the required source voltage and capacities. The implementation of this process is figured by Figure 5, 6, 7 Figure 5 describes parallel circuit or arrangement, Figure 6 describes series circuit or arrangement, Figure 7 describes series-parallel circuit or arrangement. These processes based on the concept of in series circuit the current that pass through each component that put in series is basically the same, or it can be said $i_1=i_2=i_3$; and the voltage of each component that put in series is different, or it can be said $V_1 \neq V_2 \neq V_3$; and vice versa to the parallel circuit.

3. Results and Discussion

This section discusses simulation results of each battery configuration using a mathematical model of lead-acid battery explained above. The simulation runs by using MATLAB/Simulink. Each configuration - series, parallel, and series-parallel - is done to study the behaviour of each battery configuration. Its behaviour will be figured by graphic of SoC (and $I_{battery}$) over time and battery voltage (V_{out}) over time. In this simulation, it also simulates a different number of batteries (described as n) for each configuration to show the effect of battery number to the behaviour of each configuration.
The simulation will use time sampling of 250 millisecond and the initial SoC is 1 (or 100%).

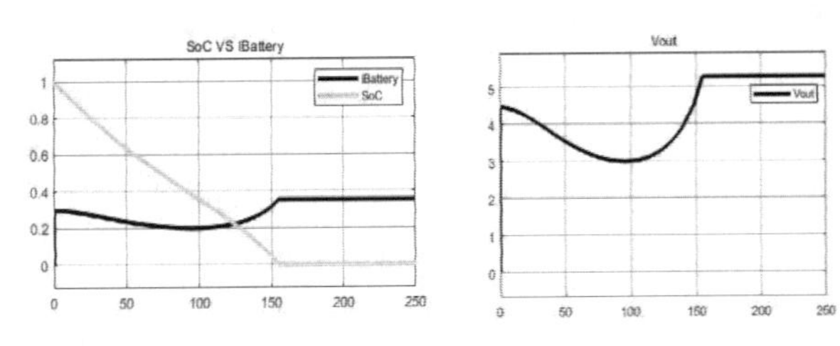

(a) (b)

Figure 8. In Series Circuit with $n = 5$ (a) SoC and $I_{battery}$ (b) V_{out}

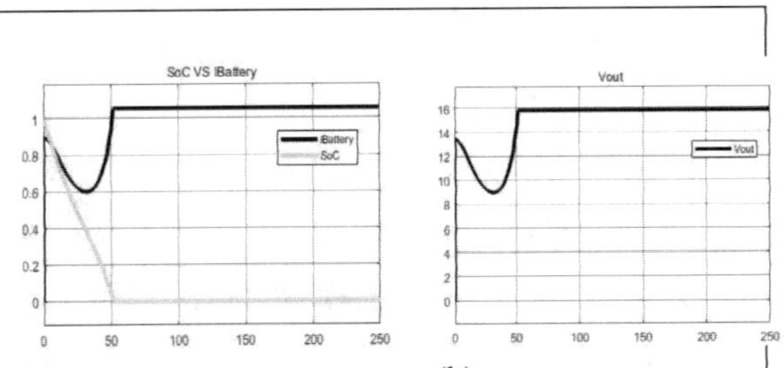

(a) (b)

Figure 9. In Series Circuit with $n = 15$ (a) SoC and $I_{battery}$ (b) V_{out}

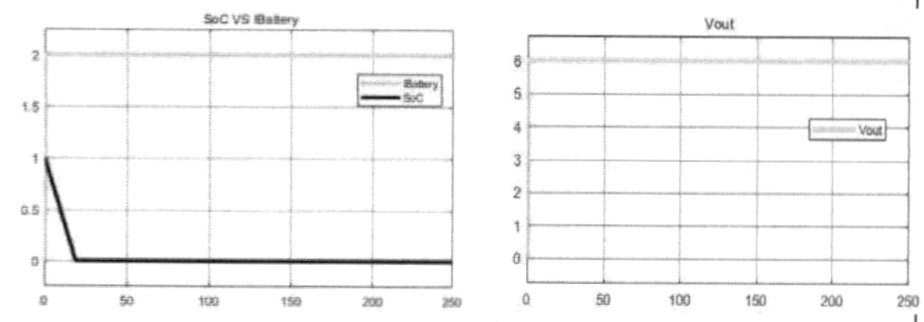

(a) (b)

Figure 10. In Parallel Circuit with n = 5 (a) SoC and $I_{battery}$ (b) V_{out}

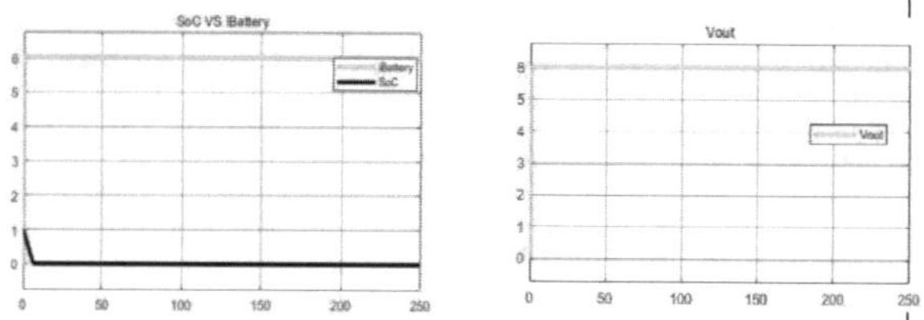

(a) (b)

Figure 11. In Parallel Circuit with n = 15 (a) So and $I_{battery}$ (b) V_{out}

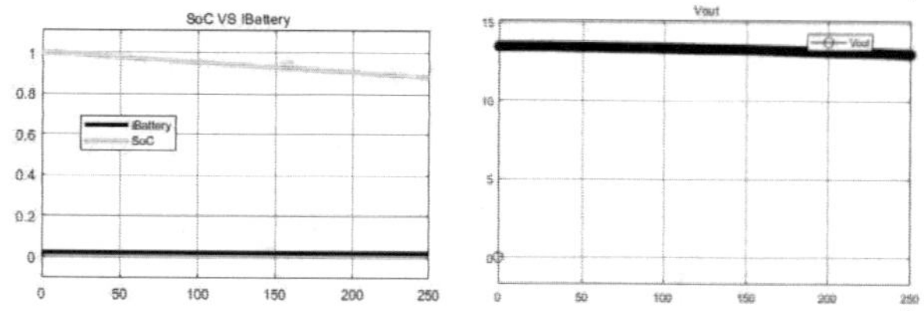

(a) (b)

Figure 12. In Parallel-Series Circuit with n = 50-15 (a) SoC and $I_{battery}$ (b) V_{out}

From figures above, SoCs will decrease slowly to 0 depending on the currents flow from the battery and if the current in simulation is increased, the SoC will be empty or going down to 0 faster as well. This behaviour is as stated in Eq. (6) that the higher the I_{batt}, the higher the reduced value of SoC, thus, it makes the SoC value down to zero faster, this phenomenon can be called a discharge process. During the discharge process, each cell shares different current due to the difference of internal resistance or polarization effects. Hence, the SoC of each cell changes with different rates and at the end of discharge, when cells together reach the lower limit (6 Volt, 0% of SoC).

The n over here is the arranged number of cells so that we can compare between series, parallel, and series-parallel circuits based on the number of n. The influence of n to the output of simulation is because the number of batteries will affect the mathematical equation of each battery configuration, thus, each n number of batteries will give different output of simulation. As the figures show, the more n-series the more voltage we get, the more n-parallel the more ampere we get, and the more n-parallel-series the more voltage and ampere we get. In [7] explained that modules can be parallel connected to obtain a higher capacity or serially connected to create battery strings with higher voltage. In turns, battery strings can be parallel connected to increase the capacity. However, from the results of simulation the optimum result is on series circuits or parallel circuits, it's slightly different with [12] the best connection topology is the series connection of parallel modules of cells provided.

4. Conclusion

From the result section, it could be seen that the behavior of lead-acid battery is divided into three kinds of type, the series circuit would make the V_{out} larger, the parallel circuit would make the $I_{battery}$ larger, and the parallel-series one would make the V_{out} and $I_{battery}$ larger. We can also conclude that, if we have an EV target with V_{out} = 400 Volt and the Capacity = 100 Ah, the one of ways we can execute is arranging batteries and make them in 400 series since n = 15 produce around 13 V so then we need that value to make it comes true.

Acknowledgement
This paper and the research behind it would not have been possible without the exceptional support of my supervisor, Feri Yusifvar. His enthusiasm, knowledge and exacting attention to detail have been an inspiration and kept my work on track our schedule and planning to final this paper.

References

[1] Jiang, J., and Zang, C. 2015. Fundamentals and Applications of Lithium-Ion Batteries in Electric Drive Vehicles. *Wiley. China*. p1
[2] Plett, L. G. 2015. Battery Management System, **1** (Artech-House). London. P 314
[3] Krivik, P. 2017. Method of SoC Determination of Lead-Acid Battery. *Journal of Energy Storage.* **15**(2018), pp. 191-195
[4] Rancilio, G., Lucas, A., Kotsakis, E., Fulli, G., Merlo, M., Delfanti, M., and Masera. 2019. Modeling a Large-Scale Battery Energy Storage System for Power Grid Application Analysis. *MDPI*, 10.3390/en12173312, p12
[5] Yusivar, F., Haratsu, H., Sato, M., Wakao, S., Kondo, K., Matsuoka, K., and Kawamatsu, T. 2002. The Modeling of Lead-Acid Battery by Frequency-Response Characteristics. *IEEJ Transactions on Fundamentals and Materials*. pp.715-721.
[6] Vellucci, F., Pede, G., Ceraolo, M., Huria, T. 2012. Electrification of off-road vehicles: examining the feasibility for the Italian market, *in EVS26 International Battery, Hybrid and Fuel Cell Electric Vehicle Symposium* (2012), pp.1–17.
[7] Baronti, F., Di Rienzo, R., Papazafiropulos, N., Roncella, R., and Saletti, R. 2014. Investigation of series-parallel connections of multi-module batteries for electrified vehicles. *In 2014 IEEE International Electric Vehicle Conference (IEVC)*, pp. 1-7. IEEE.
[8] Baronti, F., Fantechi, G., Roncella, R., Saletti, R., Pede, G., and Vellucci, F. 2013. Design of the battery management system of LiFePO 4 batteries for electric off-road vehicles. *In 2013 IEEE International Symposium on Industrial Electronics*, pp. 1-6. IEEE.
[9] Ye, M., Song, X., Xiong, R., and Sun, F. 2019. A novel dynamic performance analysis and evaluation model of series-parallel connected battery pack for electric vehicles. *IEEE Speciial Section on Advanced Energy Storage Technologies and Their Applications,* p.14256-14264
[10] Xu, J., Cao, B., and Wang, J. 2016. A novel method to balance and reconfigure series-connected battery strings. *Energies*, **9**(10), pp.766.
[11] Gong, X., Xiong, R., and Mi, C. C. 2015. Study of the Characteristic of Battery Packs in Electric Vehicles With Parallel-Connected Lithium-Ion Battery Cells. *IEEE Transactions on Industry Applications*, **51**(2), 1872-1878.

IOP Conf. Series: Materials Science and Engineering 1158 (2021) 012005 doi:10.1088/1757-899X/1158/1/012005

Supercapacitor Modeling for Conversion Systems and Power Supply in Electric Vehicles

Fadilah*, Y Feri
Electrical Engineering Department, Universitas Indonesia, Depok, Indonesia

Email: *fadilah91@ui.ac.id

Abstract. The main obstacle in an Electric Vehicle (EV) is the large torque requirement at the start. And in this case, it must be overcome with a battery that is bigger and heavier. However, a large battery will become a burden when moving and braking. The use of supercapacitors can make the vehicle move more powerful at the beginning of the movement because of its very fast energy release. This paper presents a supercapacitor modeling to support battery performance in an energy storage system. This research is focused on modeling the supercapacitor connected to the drive converter circuit by looking for a mathematical model in the form of a state-space equation, the results of which are simulated using the C Mex S function in Matlab / Simulink. From the modeling carried out, it is obtained a system that has a system response that is built and shows the system is working properly, the output voltage can follow the reference results.

1. Introduction

The development of technology will affect the development of energy storage systems, especially in electric vehicles. However, the problem that is often faced with electric vehicles is the need for large torque at the start. Thus, a storage system that can support a system such as batteries, batteries, and supercapacitors is needed [1].

Supercapacitors, also known as ultracapacitors or electric double-layer capacitors (EDLC), have unique characteristics that can be used as energy storage in various applications, one of which is a hybrid system [2]. On a life cycle basis, the supercapacitor has a longer life cycle than the battery. Besides that, another advantage of the supercapacitor is the very fast filling and discharging process, so it is very suitable for use in circuits that require a fast release of energy with a large amount of energy. The theory on which the supercapacitor is based was first described by Hermann von Helmholtz in 1853. He argued that the interaction between the conductor and the electrolyte in the capacitor is determined by the electrostatic relationship and that no chemical reaction is involved in the process [3]. One of the problems with the use of supercapacitors is how accurate the State of Charge (SOC) of the Energy storage system [4].

This paper presents two aspects. The first is supercapacitor modeling which is based on parameter values that vary according to the State of Charge (SOC). The second presents the Buck-Boost Converter model which is connected to the supercapacitor. Modeling the two proposed models, namely by looking for a mathematical model into the form of state space and simulated using CMex S-Function in Simulink Matlab [5]. This paper presents a supercapacitor modeling connected to a buck-boost converter, by looking for a mathematical model into state space and simulating it using the cmex s-function in Matlab [2]. From the results obtained, it is known that modeling using cmex has a good response and the model built has characteristics that resemble the original.

IOP Conf. Series: Materials Science and Engineering 1158 (2021) 012005 doi:10.1088/1757-899X/1158/1/012005

2. Method

Supercapacitor and Buck-Boost Converter are modeled by determining the state Space equation model. State-space modeling is an analytical method for a complex control system [6]. This method is used to analyze a control system with multiple inputs and multiple outputs or what is called Multiple Input and Multiple Output (MIMO). The State Space equation model is expressed in the form of a matrix (1) (2) where A, B, C, and D are state-space matrices, $\dot{x}$ is the derivative matrix of the x matrix, u is the input matrix and y is the output matrix.

$$\dot{x} = Ax + Bu \quad (1)$$

$$y = Cx + Du \quad (2)$$

2.1. Supercapacitor State Space Model

The Supercapacitor model described in the literature is usually based on an equivalent electrical circuit that can represent a voltage or current profile. The supercapacitor model is shown in Figure 1. In the supercapacitor circuit, the voltage across the main capacitance is OCV (Open Circuit Voltage), which corresponds to the voltage that appears at the supercapacitor terminal after the transient remains. Because the transients simulated by the R-C block have a time constant.

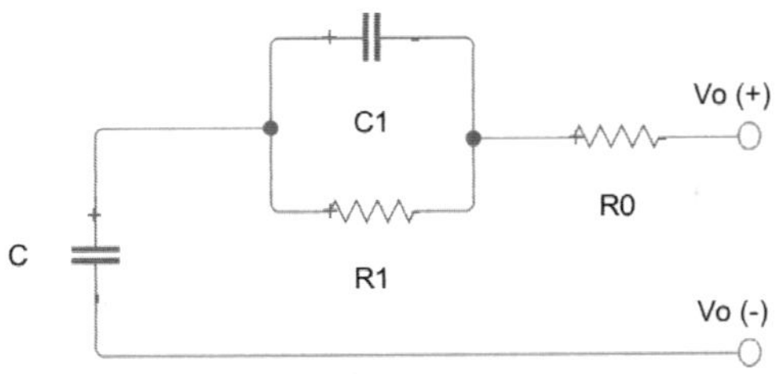

Figure 1. Model for Supercapacitor.

From the supercapacitor model that has been selected, the derivative equations (3) and (4) can be obtained. From the derivative equation, it can be written into a state-space model, with the state variable in the model determined, namely $V_c = x_1$ 1 and $V_{c1} = x_2$, with $I_o = u$. Then it can be expressed in the form of a matrix as defined in equations (1) and (2), the state space equations (5) and (6) are obtained.

$$\frac{dV_c}{dt} = -\frac{I_o}{C} \quad (3)$$

$$\frac{dV_{c1}}{dt} = -\frac{V_{c1}}{R_1 * C_1} - \frac{I_o}{C_1} \quad (4)$$

$$\begin{bmatrix} \dot{x_1} \\ \dot{x_2} \end{bmatrix} = \begin{bmatrix} 0 & 0 \\ 0 & -\frac{1}{R_1 . C_1} \end{bmatrix} \begin{bmatrix} V_c \\ V_{c1} \end{bmatrix} + \begin{bmatrix} \frac{1}{C} \\ 0 \end{bmatrix} I_o \quad (5)$$

$$\begin{bmatrix} y_1 \\ y_2 \end{bmatrix} = \begin{bmatrix} 1 & 0 \\ 0 & 1 \end{bmatrix} \begin{bmatrix} V_c \\ V_{c1} \end{bmatrix} + \begin{bmatrix} 0 \\ 0 \end{bmatrix} u \quad (6)$$

One of the most common techniques for estimating SOC is called the OCV-SOC correlation. This technique consists of a simple ampere-clock calculation, with voltage-based compensation. The supercapacitor SOC equation (7) to obtain the parameters from the SOC output value table 1. From the selected supercapacitor model Figure 1 with the characteristics of the supercapacitor table 2.

$$SOC = \frac{Q}{Q_{maks}} = \frac{\int i(t)dt}{CV_{maks}} = \frac{V_{oc}}{V_{maks}} \quad (7)$$

An experimental test has also evaluated the dynamics of the model response of the element values R0, R1, C1 at different SOC [5].

Table 1. Identification of Model Parameters as a function of SOC

SOC	$V_{oc}(V)$	$R_o(m\Omega)$	$R_1(m\Omega)$	$R_1C_1(s)$
0.52	1.43	0.51	0.11	4.2
0.76	2.04	0.50	0.10	11
1	2.85	0.48	0.09	16

Table 2. Characteristics of the Supercapacitor cell

Nominal Capacitance (F)	3000
Maximum Voltage (V)	2.85
Nominal Voltage (V)	2.7
Max Continuous Current (A) ΔT **15**°C	130
Max Continuous Current (A) ΔT **40**°C	210
Max Peak Current (A)	2000
Internal resistance (mΩ)	0.29
Leakage resistance Current* (mA)	5.2
Mass (kg)	0.5

* after 72 Hours at nominal voltage, the initial leakage current can be higher

2.2. Buck Boost Converter State Space Model

The buck-boost converter is a type of DC-DC converter that is capable of producing DC voltage variations that are smaller or lower than the input voltage according to the switching frequency [7]. The components of the buck-boost converter consist of several constituent components, namely the transistor, inductor (L), diode (D), and capacitor (C) which are arranged as shown in Figure 2.

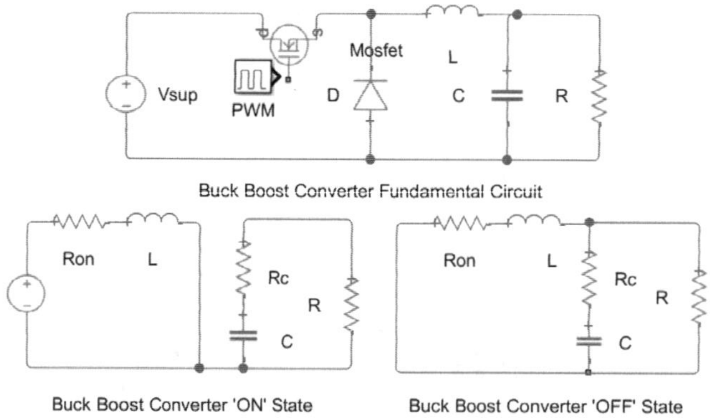

Figure 2. Buck Boost Converter Fundamental, 'On' and 'Off' state equivalent

The fundamental buck converter circuit and is On' and 'Off' state equivalent. On condition 'On', the inductor gets input from V_{sc}. For current flow to the capacitor ($I_L = 0$). Equations derived from the state

(8) and (9) with state variables namely V_c and I_L. It can be written into a matrix of state spaces A and B as defined in equation (10).

$$\frac{di_L}{dt} = \frac{V_{SC}}{L} - \frac{I_L.R_{on}}{L} \tag{8}$$

$$\frac{dV_c}{dt} = \frac{V_c}{C.R_c} \tag{9}$$

$$\begin{bmatrix} \dot{x_1} \\ \dot{x_2} \end{bmatrix} = \begin{bmatrix} 0 & -\frac{R_{on}}{L} \\ \frac{1}{R.Rc} & 0 \end{bmatrix} \begin{bmatrix} V_c \\ I_L \end{bmatrix} + \begin{bmatrix} \frac{1}{L} \\ 0 \end{bmatrix} u \tag{10}$$

In the 'Off' state, the voltage source is zero. Then the derivative equation obtained (11) and (12)

$$\frac{di_L}{dt} = \frac{I_L.R_{on}}{L} \tag{11}$$

$$\frac{dV_c}{dt} = -\frac{V_c}{(R_c + R)C} + \frac{I_L.R}{(R_c + R)C} \tag{12}$$

From equations (11) and (12), the state variable $V_c = x_1$ 1 and $I_L = x_2$ then we can write equation (13) state space model for the 'Off' condition of the Buck Boost Converter circuit.

$$\begin{bmatrix} \dot{x_1} \\ \dot{x_2} \end{bmatrix} = \begin{bmatrix} -\frac{1}{(R_c + R)C} & \frac{R}{(R_c + R)C} \\ 0 & \frac{R_{on}}{L} \end{bmatrix} \begin{bmatrix} V_c \\ I_L \end{bmatrix} + \begin{bmatrix} 0 \\ 0 \end{bmatrix} u \tag{13}$$

Matrix C and the matrix D in the state space model can be written in equation (14).

$$\begin{bmatrix} y_1 \\ y_2 \end{bmatrix} = \begin{bmatrix} 1 & 0 \\ 0 & 1 \end{bmatrix} \begin{bmatrix} V_c \\ I_L \end{bmatrix} + \begin{bmatrix} 0 \\ 0 \end{bmatrix} u \tag{14}$$

After obtaining the A and B matrices in the 'On' and 'Off' state in the buck boost converter circuit. A matrix of the mean space equations (15) and (16) is required. The equilibrium state is the superposition of the state space matrix of the two conditions, taking into account the duty cycle (d), thus obtained (18) and (19).

$$\dot{x} = Ax + Bu \tag{15}$$

$$y = C^T x \tag{16}$$

Where,

$$A = dA_{ON} + (1 - d)A_{OFF} \tag{17}$$

$$A = d\begin{bmatrix} 0 & -\frac{R_{on}}{L} \\ \frac{1}{R.R_c} & 0 \end{bmatrix} + (1-d)\begin{bmatrix} -\frac{1}{(R_c+R)C} & \frac{R}{(R_c+R)C} \\ 0 & \frac{R_{on}}{L} \end{bmatrix} = \begin{bmatrix} -\frac{(1-d)}{(R_c+R)C} & -\frac{R_{on}.d}{L} + \frac{R\,(1-d)}{(R_c+R)C} \\ \frac{d}{R.R_c} & \frac{R_{on}(1-D)}{L} \end{bmatrix} \quad (18)$$

$$B = dB_{ON} + (1-d)B_{OFF} \quad (19)$$

$$B = d\begin{bmatrix} \frac{1}{L} \\ 0 \end{bmatrix} + (1-d)\begin{bmatrix} 0 \\ 0 \end{bmatrix} = \begin{bmatrix} \frac{d}{L} \\ 0 \end{bmatrix} \quad (20)$$

To complete the buck boost converter model, the average state space matrix, can be substituted for equations (1) and (2), then the state space equation of the buck boost converter system becomes (18).

$$\begin{bmatrix} \dot{x_1} \\ \dot{x_2} \end{bmatrix} = \begin{bmatrix} -\frac{(1-d)}{(R_c+R)C} & -\frac{R_{on}.d}{L} + \frac{R\,(1-d)}{(R_c+R)C} \\ \frac{d}{R.R_c} & \frac{R_{on}(1-D)}{L} \end{bmatrix} \begin{bmatrix} V_c \\ I_L \end{bmatrix} + \begin{bmatrix} \frac{d}{L} \\ 0 \end{bmatrix} u \quad (21)$$

3. Results and Discussion

From the supercapacitor model and the buck boost converter model, the state space equation for each condition has been obtained. These equations are simulated using the S-Function block in Matlab Simulink [5]. The block diagram of the simulation system is shown in Figure 3. It consists of a Supercapacitor block, a Buck Boost Converter block, PID controller, and PWM generator.

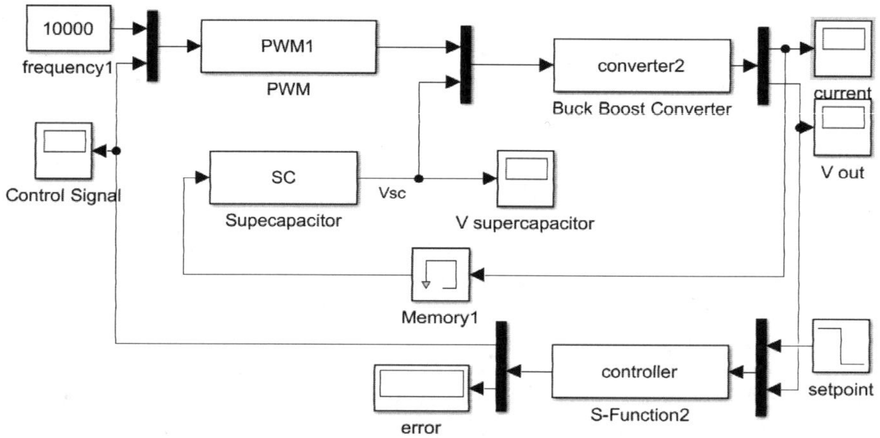

Figure 3. Simulated Block Diagram with Cmex S-Function in Matlab Simulink

The Pulse Width Modulator (PWM) block converts the output from the compensator to the duty ratio. The output voltage of the compensator V_c is compared to a triangular wave with an amplitude of V_p. The output of the PWM block is the high value when V_c is greater than the triangle wave and is zero when V_c is smaller than the triangle wave. If the output voltage V is smaller than the output voltage V_{ref}, the error value will increase so that the V_c value increases and the duty ratio increases. Conversely, if the large output voltage V increases, it will reduce the duty ratio. This PWM pulse will then be used to open and close the switches on the switching converter.

The parameter values of the supercapacitor are obtained based on SOC (Table 1) and the boost converter is simulated with the parameter values $R = 1{,}5\,\Omega, C = 10^{-3}F, L = 10{,}3 \times 10^{-3}H$, the switching frequency is 10 kHz, the duty cycle is 25%, the buck boost converter voltage source is obtained from the output voltage of the supercapacitor model of 2.851 V. The output is voltage and current, where the output current is used as the input of the supercapacitor, the simulation results are shown in Figure 4. From the whole system, the output value of the V_c converter is 0.007114V, the error value is obtained from the comparison of the set point with the V_c output value of -0.007115 V, the duty cycle of 0.03729 which is located in the control section with a value of K_p = 10-1, $T_i = 4e^{-3}$ and T_d = 0, by simulating the reference voltage is set from 1 V then at 5 seconds it drops to 0.25 V Figure 4. The output of the duty cycle is inputted to drive the PWM. Looks like the system is working properly. Where the output voltage can follow the reference.

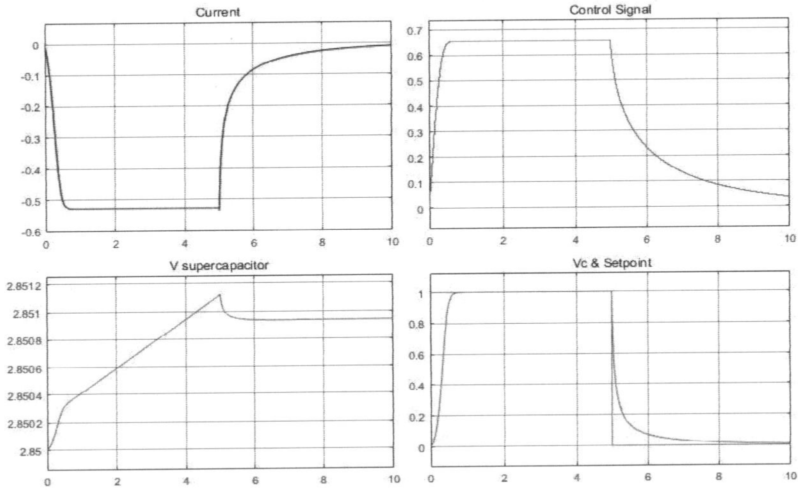

Figure 4. Simulation output results from the Buck Boost Converter circuit

4. Conclusion

From the simulation that has been done with Cmex S-Function on Simulink/Matlab. Then the final result shows that the system built is in accordance with its characteristics and has a response that resembles the original and the system can work well. Where the output voltage can follow the reference.

Acknowledgment

The authors would like to thank all parties involved in this research so that we can complete this research.

References

[1] Devillers N., Jemei S., Péra MC., Bienaimé D & Gustin F. 2014. Review of characterization methods for supercapacitor modelling. *Journal of Power Sources*, **246,** pp. 596–608. Elsevier.

[2] Yusivar F & Wakao S. 2001. Minimum requirements of motor vector control modeling and simulation utilizing C MEX S-function in Matlab/Simulink, *Proc. IEEE PEDS*, pp. 315-321.

[3] Belhachemi F., Rael S, and Davat B. 2000. A physical based model of power electric double-layer supercapacitors, *Conf. Rec. IEEE Ind. Appl. Conf. Thirty-Fifth IAS Annu. Meet. World Conf. Ind. Appl. Electr. Energy* (Cat. No.00CH37129), **5**, pp. 3069–3076. IEEE.

[4] Camara MB & Hamid GF. 2010. DC/DC Converter Design for Supercapacitor and Battery Power Management in Hybrid Vehicle Applications—Polynomial Control Strategy. *TRANSACTIONS ON INDUSTRIAL ELECTRONICS*, **57**(2), pp 587-597. IEEE.

[5] Ceraolo M., Lutzemberger G and Poli D. 2017. state-Of-Charge Evaluation Of Supercapacitor. *Journal of Energy Storage,* **11**, pp. 211-218. Elsevier.

[6] Lingling D. 2009. Study on supercapacitor equivalent circuit model for power electronics applications, *Power Electronics and Intelligent Transportation System International Conference on (PEITS)*, **2**, pp 51–54.
[7] Rodney HGT & Landon YHH. 2015. DC-DC Converter Modeling and Simulation using state space Approach, pp 42-47. IEEE.
[8] Popoola K. 2018. Modelling and Simulation of Supercapacitor for Energy Storage Applications. *ADBU – Journal of Engineering Technology*. Vol. **7**, pp 1-6.
[9] Yujie W., Chang L., Rui P & Zonghai C. 2017. Modeling and State-of-Charge prediction of Lithium-ion battery and Ultracapacitor hybrids with a co-estimator. *Energy*, **121**, pp 739-750.
[10] Zhang L., Xiang L., Wang Z., Sun F & Dorell DG. 2018. A review of supercapacitor modeling, estimation, and aplications: a control management perspective. *Renewable and Sustainable Energy Reviews*, **81**, pp 1868-1878. Elsevier.

The Influence of Magnet Number and Dimension on Torque Characteristics in the Interior Permanent Magnet Synchronous Motor (PMSM)

K Yulianto[1*], F Yusivar[2], B Sudiarto[3]**
[1,2,3] Department of Electrical Engineering, Universitas Indonesia, Depok 16424, Indonesia

Email: *katri.yulianto2@gmail.com, **yusivar@yahoo.com

Abstract. Permanent magnet synchronous motors (PMSMs) are commonly used in electric vehicles because they have specific advantages such as high power density, high torque, and high efficiency. This paper presents the design of an interior permanent magnet (IPM) rotor motor with 5kw capacity as an electric motorcycle propulsion machine. Principal design calculations obtain initial design parameters. The choice of using a magnet dimension has a significant impact on performance. Optimization is applied to achieve the required torque characteristics by varying the permanent magnet size with fixed volume and length. The simulation is performed using finite element analysis software to obtain the speed and torque characteristics. The results showed that the motor torque was higher with a wider magnet scale and fewer magnets per pole. Furthermore, using thicker magnets obtained better motor efficiency. Hence, this needs to be taken into consideration when designing the motor according to torque specifications.

1. Introduction

The development of transport technology attempts to minimize the consumption of fossil fuels, one of which is electric vehicles. One of the main components used as propulsion in the electric vehicle is an electric motor. Permanent magnet synchronous motors (PMSMs) are currently widely used in electric vehicles because they have several advantages: high power density, reduced weight and volumetric dimensions, high torque, and high efficiency [1]. Several parameters may have an impact on performance in the design of PMSMs, one of which is the use of permanent magnets. PMSMs characteristics are strongly influenced by the characteristics and size of the permanent magnets. It is important to know the application of the motor [2].

The selection of the dimensions of permanent magnets in the brushless motor will significantly affect motor performances. Several findings on the effect of magnetic size changes have been obtained from previous research. Maximum output torque will be achieved when both the length and width of the magnet are maximized. The increment of torque will not determine the increases in motor efficiency as well [3]. According to magnet size, air gap clearance changes and torque of motor changes. Air gap clearance is proportional to permanent magnet width. If air gap clearance is long, the flux line and torque are reduced even though the permanent magnet usage increases [4]. Modifying the width of the stator teeth, geometry, and the width of the permanent magnets on the rotor is the most efficient way to reduce the cogging torque [5,6]. The effect of the number of pole pairs and the width of the magnet on the mechanical torque of magnetic coupling is obtained. As wider are the permanent magnets, higher is the mechanical torque of the magnetic coupling [7]. Based on the findings of previous research, this paper presents the PMSM design process by varying the size and the number of magnets per pole in the rotor's interior permanent magnet.

The purpose of this research is to design a PMSM 5 kW capacity as an electric motorcycle propulsion machine. The design target is torque close to 40 Nm at rated speed. The design considers the availability of permanent magnet sizes in the laboratory. Principal design calculations obtain initial design parameters. The design was simulated using finite element analysis software to achieve performance. The choice of using a magnet dimension has a significant impact on the performances. Optimization was applied to achieve the required torque characteristics by varying the number of permanent magnets per pole and magnet size with fixed volume and length. Therefore, the efficient design of the number and dimensions of magnets is expected to minimize costs.

2. Method

2.1. Process design

Several processes are performed to design a PMSM. The stages of the process design flow diagram can be seen in Figure 1.

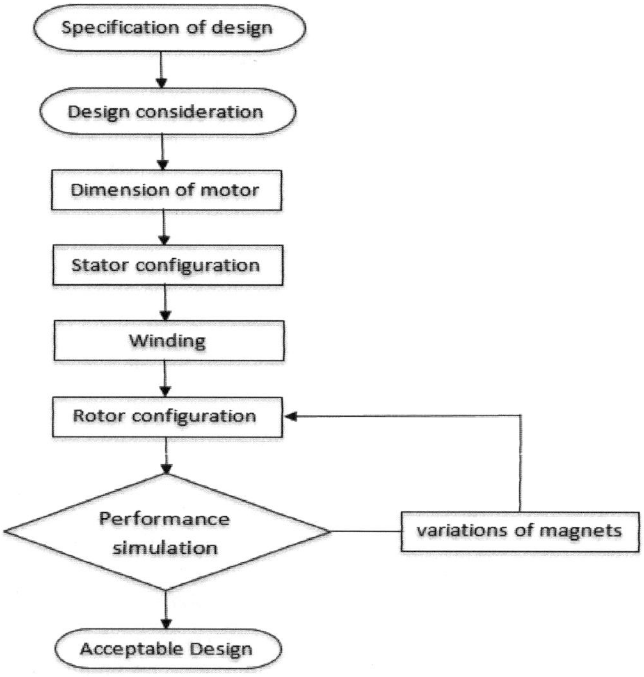

Figure 1. Process design flow diagram

The design begins with the specification of the design. Design considerations are interior permanent magnet type rotor and M250-35A used for stator and rotor laminations material, power output (Pout) 5 kW, 1200 rpm synchronous speed, 100 Hz frequency, 3-phase star supply voltage, and 72 V DC-link voltage on inverter side. The inverter voltage drop is assumed at 5% [8]. The NdFeB N48 permanent magnets should be distributed at the interior permanent magnet rotor. The length of the permanent magnet is 40 mm. The product $\eta Cos\ \emptyset$ the rated load is assumed at 0.9.

2.2. Principle of Calculation

There are several calculation procedures, and the following are the motor calculation steps [9-11].

The rated armature current is calculated by following equation (1).

$$Ia = \frac{Pout}{3V_l\, \eta\ Cos\ \emptyset} \qquad (1)$$

V_l is the inverter's output voltage. Its value takes into account the voltage-drop.

The number pole pairs,

$$p = \frac{f}{n_s} \tag{2}$$

Where *f* is frequency and *ns* is rotating speed in rotating per second (rps). The number of poles is eight for rated speed 1200 rpm.

Output coefficient,

$$\sigma p = \frac{Pout\ \epsilon}{D_{lin}{}^2 L_i n_s} = 0.5\,\pi^2\, k_{w1}\, A_m\, B_g\, \eta\, Cos\ \emptyset \tag{3}$$

Here, k_{w1} is stator winding factor. The value is between 0.8 - 0.99. A_m is stator line current density. A_m ranges from 10000A/m for small motors to 55000A/m for medium power motors. This design has been assumed to be 45000A/m. ϵ is no load EMF to phase voltage ratio (0.60 – 0.95). From equation (3), we get ratio stacking length L_i and stator inner diameter D_{lin}. The design is defined using a laminate length of 40 mm. It is due to the availability of magnets in our laboratory. The ratio of the stator's inner diameter to the length of the stacking is obtained.

Outer stator slot diameter,

$$D = \frac{Do - 0.647}{1.175 + \frac{1.03}{p}} \tag{4}$$

Width teeth body,

$$w_{tb} = \frac{2\pi\, R_{ro}\, B_g}{N_s\, K_{st}\, B_t} \tag{5}$$

Width stator yoke,

$$w_{sy} = \frac{\pi\, R_{ro}\, B_g}{N_m\, K_{st}\, B_{sy}} \tag{6}$$

Where Rro is outer rotor diameter, Ns is number slot stator, Kst is lamination stacking factor (0.8-0.99), Bg is air gap flux density, for NdFeB N48 = 1.38 T. B_t is stator teeth density and B_{sy} is stator yoke flux density.

Diameter frame,

$$D_f = D_o + 2\, w_{sy} \tag{7}$$

The cross-section area of the armature conductors is

$$sa = 2\, \frac{\pi da^2}{4} \tag{8}$$

Three parallel conductors AWG 12 with 3.31mm^2 have been chosen. The cross-section area of the armature conductor is 9.93mm^2.

The current density in the armature winding,

$$J_a = \frac{I_a}{S_a} \tag{9}$$

The current density is 7.0 A/mm^2. This value can be accepted for continuous operation of small PMSM motors.

Pole pitch,

$$\tau = \frac{\pi D_{lin}}{2p} \tag{10}$$

Coefficient volume,

$$C_v = \frac{2\, k_{ocf}\, k_f\, k_{ad}\,(1+\epsilon)}{\pi^2 \xi} \tag{11}$$

The overload capacity factor has been assumed k_{ocf}= 1.5, the coefficient utilization of the permanent magnet can be estimated as $\xi = 0.5$. k_f is the form factor of the excitation field, k_{ad} is the armature reaction factor in the d axis.

The volume of all PMs used in motor,

$$V_m = \frac{c_v P_{out}}{f B_r H_c} \qquad (12)$$

Br is remanent magnetic flux density, the value for NdFeb is between 0.6 – 0.8.

$$Vm = 2ph_M w_M l_M \qquad (13)$$

2p is the number of poles, h_M, w_M, l_M are the height, width and length of the permanent magnet.

2.3. Variation

The parameters obtained from the calculation will be optimized with variation of number magnet per pole, width and thickness. A variation is performed with fixed volume and length. Variation design is shown in Table 1.

Table 1. Magnet dimension of variation (w_m x t_m) in mm

	1 magnet/pole	2 magnets/pole	3 magnets/pole
Design 1	30 x 4	15 x 4	10 x 4
Design 2	26 x 4.5	13 x 4.5	9 x 4.5
Design 3	24 x 5	12 x 5	8 x 5

Determination of the magnetic variation based on the volume resulting from the calculation. The total volume of the magnet can be obtained from equation 12. The total volume of magnet used is approximately 4811.15 mm3 per pole. With a fixed magnetic volume and a lamination stacking length, we can determine the magnets' variations.

3. Results and Discussion

Parameters PMSM have been found after performing calculations. The motor parameters were shown in Table 2.

Table 2. Parameters of PMSM

Item	Value
Stator	
Outer Diameter	182 mm
Inner diameter	130 mm
Number slot	24
Height slot	16.95 mm
Width teeth body	6.22 mm
Width stator yoke	9.12 mm
Rotor	
Outer diameter	128 mm
Inner diameter	35 mm
Number of pole	8
Winding	
Number of turn	4
AWG size	12
Strand	3

The magnetic flux distribution of the PMSM was designed to be shown in figure 2.

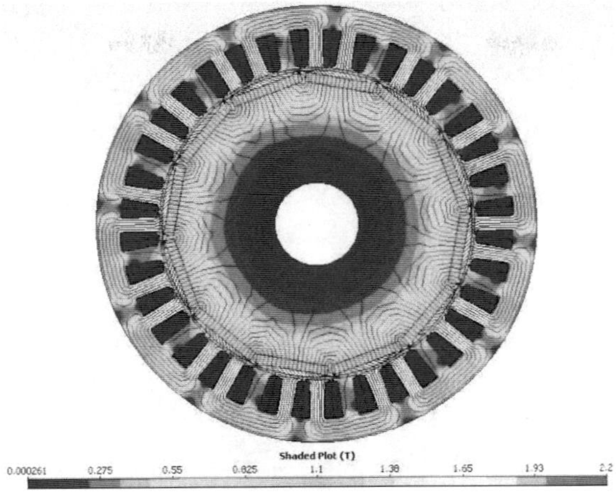

Figure 2. Magnetic flux distribution

The maximum saturation value of the lamination material used is above 1.8 T. The distribution area of the magnetic field that exceeds more than 1.8 T moves from yellow to orange. Between the corner magnet rotor and the stator slot shoe, an orange colour appears, but it's not dominant. Magnetic flux leakage causes this. By using more magnets per pole, the more likely the magnetic flux leakage will be to occur. It can be avoided by creating a flux barrier.

The PMSMs can work in the overspeed range. This means that the real speed of the motor can exceed the nominal rated speed. The torque is inversely proportional to the rpm so that the output power remains constant for the range where the nominal rpm is exceeded [12]. The ideal characteristic torque curves are shown in Figure 3.

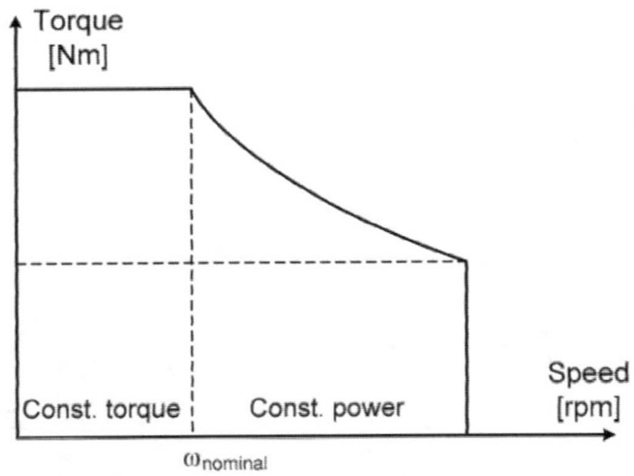

Figure 3. The ideal characteristic torque curves of PMSM

Curves of the relation between a characteristic torque and motor speed in rotations per minute are shown in Figure 4. All of which are the result of changes in the dimensions and number of magnets.

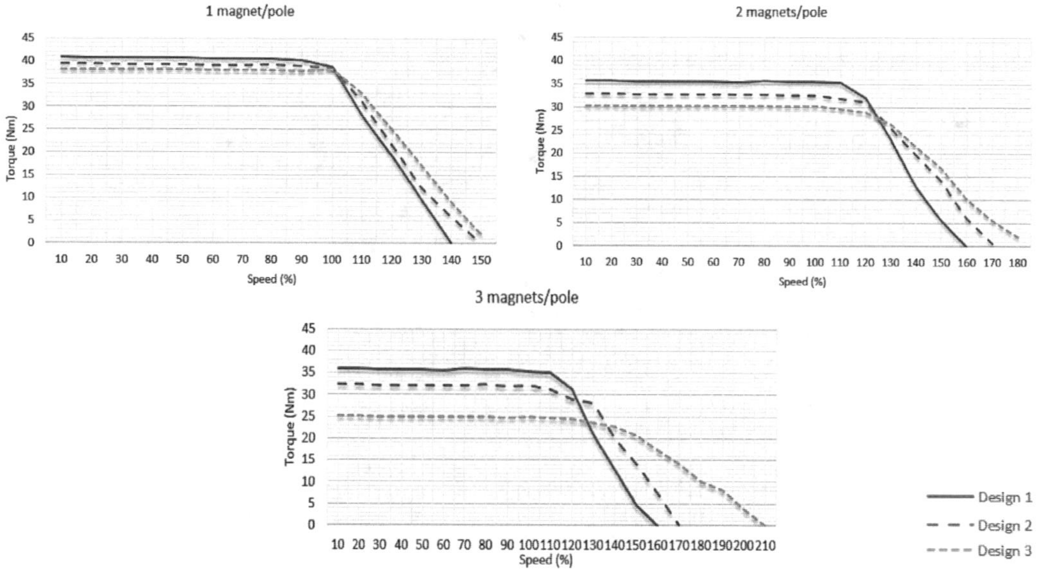

Figure 4. Torque vs speed characteristic from design variations

Based on the simulation results shown in figure 4, the highest torque was obtained by using a magnet dimension of 30x4 mm, with the number of magnets per pole being a magnet. The output torque is 38.52 Nm when the speed is 1200 rpm, but the efficiency is lower when the speed is higher. At a speed of 1680 rpm (140%), the torque becomes zero. The lowest torque was obtained using a magnet dimension of 8×5 mm with the number of magnets per pole being three magnets. The output torque is 25.1 Nm when the speed is 1200 rpm, but the efficiency is higher when the speed is higher. At a speed of 2520 rpm (210%), the torque becomes zero.

In reference [3], the output cogging torque is increased due to expansion on both parameters for height of the magnet and width of the magnet. The cogging torque of the design variation is shown in Figure 5.

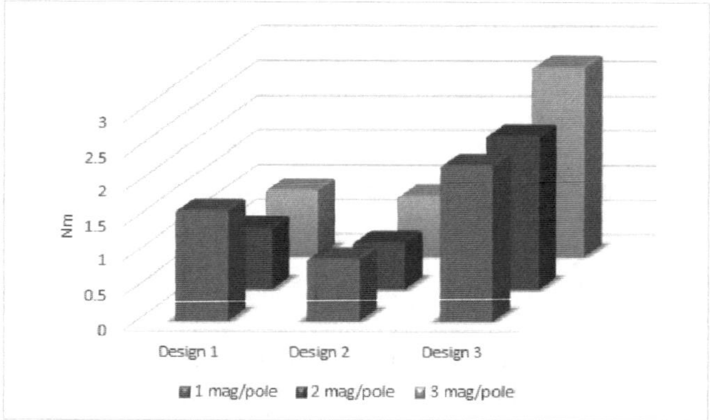

Figure 5. Maximum cogging torque from design variations

The cogging torque increases when the thickness and width of the magnets increase and the designs have the same number of magnets per pole. In design 1, the cogging torque is larger than design 2 because the width of the magnet is larger. The cogging torque in design 3 is higher than design 2 because the size of the magnet is thicker. When comparing the torque results between the number of magnets per pole, the results show that design 2, which uses two magnets per pole, has a smallest cogging torque (0.68Nm). Design 3, which uses three magnets per pole, has the largest cogging torque (2.73 Nm).

4. Conclusion

This paper provides the concept design of PMSM, Interior Permanent magnet (IPM) type rotor, with 5 kW capacity as an electric motorcycle propulsion machine. The specifications and considerations of the design are predefined. The results of the calculations and simulations were obtained in accordance with the specification requirements. The effect of the number and size of magnets per pole on motor torque and speed was discussed and analyzed. Using fewer magnets per pole and a wider scale would provide higher torque. Furthermore, using thicker magnets increases efficiency. The torque becomes zero at higher speeds. The cogging torque increases with an increase in the thickness or width of the permanent magnet.

Acknowledgment

We would like to thank the lecturers and laboratory practitioners at the Electrical Engineering Department, University of Indonesia, as well as those who contributed to this research process.

References

[1] Ahmed, A., Majed, A., Naif, F. A. and Fawas, R. 2017. Comparative Study for Machine Candidates for High Speed Traction Applications. In *International Journal of Electrical Engineering*. **10**(1), pp 71-84.

[2] Todorov, G. and Stoev, B. 2015. Analytical Model for Sizing the Magnets of Permanent Magnet Synchronous Machines. *Journal of Electrical Engineering*. **3**, pp 134-141.

[3] Norhisam, M., Nazifah, A., Aris, I., Firdaus, R. N., Wakiwaka, H and Nirei, M. 2010. Effect of Magnet Size on Torque Characteristic of Three Phase Permanent Magnet Brushless DC Motor, In *IEEE Student Conference on Research and Development (SCOReD),* pp. 293-296.

[4] Yoo, J. S., et al. 2018. Analysis of Optimize Designing Small Size BLDC Motor Considering Air Gap Clearance, In *2018 21st International Conference on Electrical Machines and Systems (ICEMS)*, pp. 265-268.

[5] Diga, N., Ghita, C., Diga, S., Constantin, D. and Stoica, C. 2015. Some Aspects of the Cogging Torque Study of a Permanent Magnet Synchronous Motor to Drive a Bicycle. In *9th International Symposium on Advanced Topics in Electrical Engineering (ATEE)*. pp.844-847.

[6] Ferkova, Z. 2015. Influence of the Arrangement and Sizes of Magnets on the Cogging Torque of PMSM. in *Technical Transactions: Electrical Engineering*. **1**, pp.161-171.

[7] Ose, B. and Pugachov, V. 2011. The Influence of Pole Pair Number and Magnets' Width on Mechanical Torque of Magnetic Coupler with Rounded Permanent Magnets. In *Scientific Journal of Riga Technical University*, pp 63-66.

[8] Mukherjee, P. and Sengupta, M. 2014. Design Analysis and Fabrication of a Brushless DC Motor. In *IEEE International Conference on Power Electronics, Drives and Energy Systems (PEDES)*, pp. 1-6.

[9] Gieras, J. F. 2010. *Permanent Magnet Motor Technology* Third Edition (New York - Boca Raton CRC Press) pp. 200-217.

[10] Hanselman, D. 2003. *Brushless Permanent Magnet Motor Design*: Second Edition. (Ohio - Magna Physics Publishing), pp. 203-206.

[11] Suryandi, A. A., Alinandar, C. S., Mandasari, D. R. and Yulianto, K. 2018. A 250 kW Three Phase Induction Motor Design for Electric Bow Thruster. In *International Conference on Electrical Engineering and Computer Science (ICECOS)*, pp. 247-252.

[12] Rudnicki, T., Czerwinski, R., Polok, D. and Sikora, A. 2015. Performance Analysis of a PMSM Drive with Torque and Speed Control. In *Proceedings of the 22nd International Conference "Mixed Design of Integrated Circuits and Systems"*, pp.562-566.

IOP Conf. Series: Materials Science and Engineering 1158 (2021) 012007 doi:10.1088/1757-899X/1158/1/012007

The Effect of 12 Pulse Converter Input Transformer Configuration on Harmonics of Input Current

Hasan Surya
Electro Engineering Department, Politeknik Negeri Bandung
Jalan Gegerkalong Hilir, Ds. Ciwaruga, Kab. Bandung Barat, Indonesia

Ehasan.surya@polban.ac.id

Abstract. This paper discusses the effect of the 12-pulse converter input transformer configuration on the input harmonic currents. This is done because there are many transformer connection configurations, which according to the standard there are 5 transformer connection groups. The research method begins with a literature study on the 12-pulse converter. Next, determine the specifications of the 12-pulse converter and build the 12-pulse converter. Then, a theoretical analysis was carried out on the waveform of the 12-pulse converter current with the Dy11/d0 and Yd1/y0 input transformer configurations. From the analysis and simulation results, it was found that the THDi of the two 6 pulse converters obtained 29% THDi, while the 12-pulse converter inputs side obtained 11.59%-13,6% THDi. The input transformer configuration does not affect the THDi results, except for the power sent by the 12-pulse converter with the input transformer configuration Yd1_y0 to 1/3 compared to the Dy11_d0 connection.

1. Introduction

The use of semiconductor power in electric power system technology continues to grow, which affects electrical loads, from domestic electrical loads to industrial electrical loads. These electrical loads are non-linear loads that produce harmonic current distortion. It causes additional losses, so that the system temperature increases, the system power factor decreases. To reduce this harmonic distortion in high power industrial loads such as high-power drive systems a 12-pulse converter is used. As stated in [1,2] this converter can be constructed by connecting in series two 6 pulse converters, each of which has a 30o phase difference. This phase shift can be obtained by configuring the input transformer connection of the converter.

This input transformer configuration will affect the harmonic content of the input current, so that it also affects the input voltage. As expressed in [1], where the harmonic input current of 6 pulses of order 5 drops to 0.03% and the voltage drops to 8%, so that the total harmonic value. The distortion also decreased, [2] where the THD of the current fell from 4.16 to 0.52%, while the THD of the voltage fell from 11.29% to 1.29%. Likewise, [3,4] stated that the 12-pulse converter using the input transformer autotransformer harmonic input currents of the order 5, 7, and 11 could be suppressed. This research was conducted to analyze the behavior of the harmonic distortion in a 12-pulse converter with a certain transformer connection. This is done because of the many types of 3 Phase Transformer connections, some three-phase transformer connections include Yy0, Dd0, Yd1, Dy1, Yd5, Dy5, Yy6, Dd6, Yd11, and Dy11.

The methodology used in this research is a quantitative method which is carried out by building a theory related to the 12-pulse converter, analyzing the behavior of harmonic currents, performing simulations with PSIM software, and measuring. It is started with determining the 12-pulse converter specifications to be researched, namely:

Input voltage: 400 Volts
Output Voltage: 150 Volt
Output Current: 10 Ampere
Input transformer configuration: Dy11/d0 and Yd1/y0

From this design, each 6-pulse converter must produce an output voltage of 75 V so that the 6-pulse converter input voltage can be derived as follows:

DC output voltage 6 pulses converter: $V_{DCout} = \frac{3.V_m.\sqrt{3}}{\pi}$, so that with a DC output voltage of 75 V, the secondary effective voltage of the star connected transformer is = 32 volts and the secondary winding of the transformer connected to the delta is = 55.5 volts.

The converter below shows a 6 pulse converter with a Dy11 input transformer connection and has a waveform of output voltage, output current, and input current as shown in Figure 1.

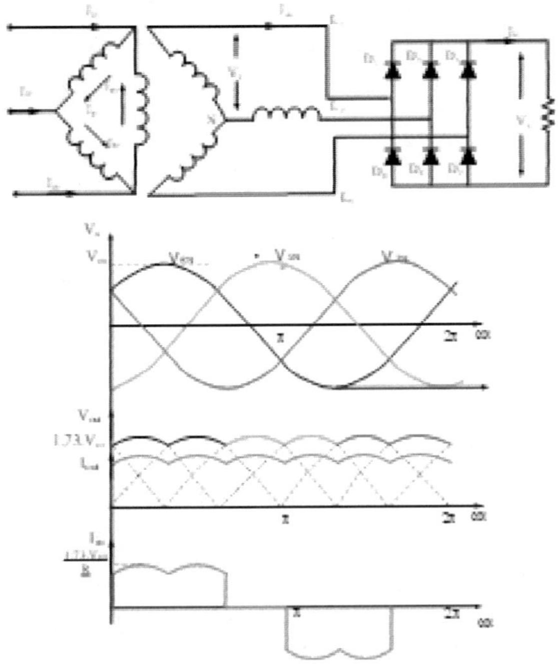

Figure 1. 6 Pulse converter, output waveform, and input current with Dy11 input transformer

This current flow in the y phase winding. Using the approximation that the waveform is a square wave. The current is in the phase D winding. In phase with the current in the y phase winding and the source current to the D winding has a form that is the result of subtracting phase 1 and phase 3 (see the current loop in circuit schematic). The results of which are shown in Figure 2.

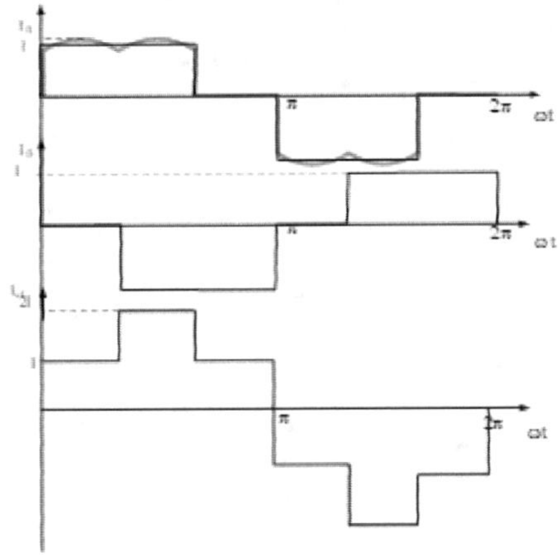

Figure 2. Waveform of the input current for the Dy11 transformer connection

IOP Conf. Series: Materials Science and Engineering 1158 (2021) 012007 doi:10.1088/1757-899X/1158/1/012007

The following is a 6 pulse converter with input transformer connection Yd1, where the primary winding has a Y connection, while the secondary winding has a Delta connection. With a delta connection, line currents and phase currents have a phase difference of 30°. The 6 pulse converter that is supplied by a transformer with a Yd1 connection will have a voltage wave and an output-input current as shown in Figure 3.

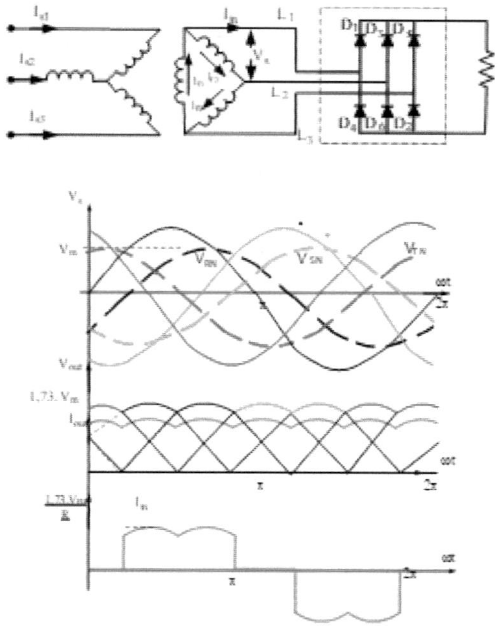

Figure 3. 6 Pulse converter, output waveform and input current with Yd1 input transformer

This current flow in the Line input converter, where the line current is the result of the reduction of the phase 1 current with the 3 phase current in the Delta winding so that the line current value is shifted 30° from the phase current. Using the approximation that the waveform is a square wave. It appears that the input current waveform in the phase D winding, the result is shown in Figure 4. The input current in the phase D winding is in phase with the current in the phase w winding, yang. So that the input current waveform on the converter is the same, both on the Dy11 and Yd1 connections. The effect of the input transformer connection only affects the phase position of the relevant input current.

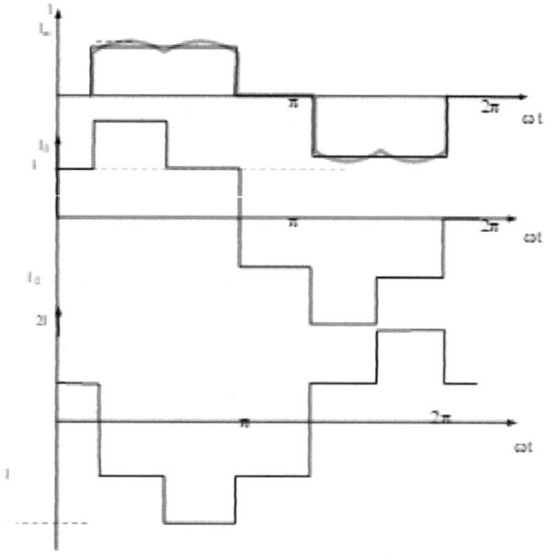

Figure 4. Waveform of Yd1 transformer connection input current

As expressed in [1], where the harmonic input current of the converter 6 pulses of order 5 drops to 0.03% and the voltage drops to 8%, so that the total harmonic distortion value also decreases, [2] where the current THD drops from 4.16 to 0,52% while the THD Voltage decreased from 11.29% to 1.29%.

2. Methodology

12 PULSE CONVERTERS

A 12-pulse converter is a converter that converts AC voltage into DC voltage, wherein one cycle the AC voltage will be converted into a 12-pulse DC voltage. In [3], [4] states that the 12 pulse converter using the autotransformer input transformer harmonic input current order 5, 7, and 11 can be suppressed. In addition to using a passive filter to reduce the input current harmonics to the converter, the transformer configuration used can also reduce the input current harmonics [5,6,7]. In addition, to have an impact on the harmonic content of the use of an input transformer, it also affects the capacity of the transformer used [3] and [5], [6] which states that the required transformer capacity is only about 40% of the DC load power. Also, in [7] shows that the input transformer connection can also be used to increase the power factor naturally. 12-pulse converter with autotransformer input transformer as stated in [8] with a zigzag connection can improve the quality of electrical power.

The 12-pulse converter circuit with input transformer configuration Dy11 / d0 is shown in Figure 5. Assuming the load has an induction of 0.1 H and R = 15 Ohm, then the waveform of the input current in each part of the converter is a square wave as shown in Figure 6.

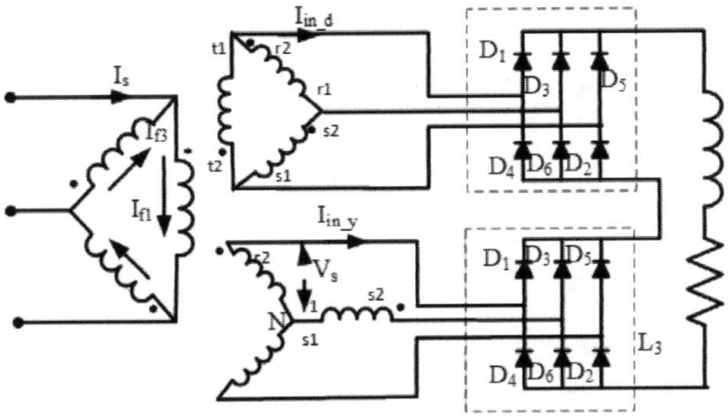

Figure 5. 12-pulse converter with input transformer Dy11 / d0 connection

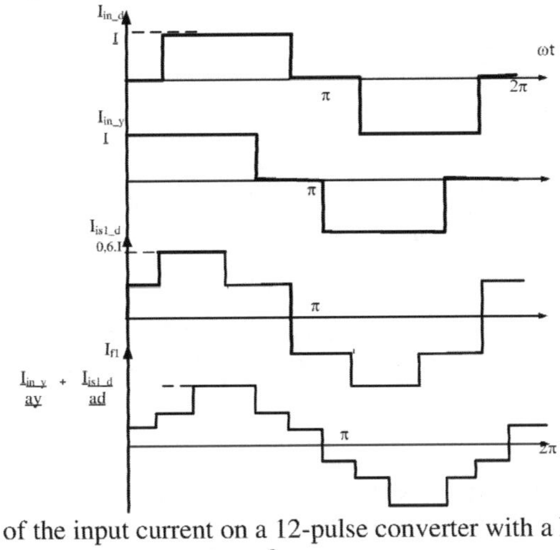

Figure 6. The Waveform of the input current on a 12-pulse converter with a Dy11 / d0 connection input transformer

The 12-pulse converter circuit with the input transformer configuration Yy0 / d1 is shown in Figure 7, with the input current waveform in each part of the converter is a square wave as shown in Figure 8. Because using the same transformer, the input voltage for each phase will be 0.58 from the previous configuration so that the output voltage becomes 87 volts, and the output current becomes 5.8 Ampere.

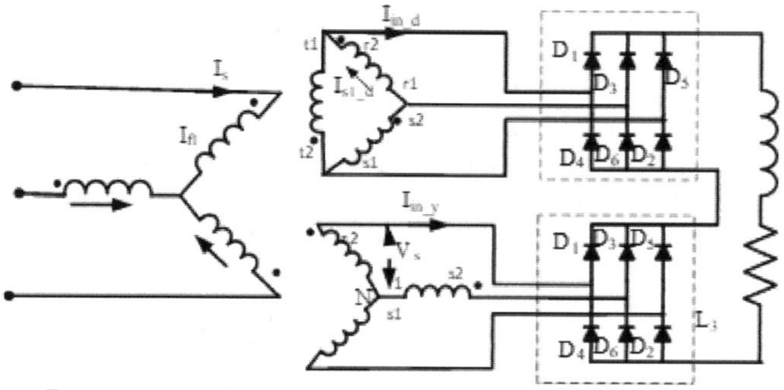

Figure 7. 12-pulse converter with input transformer configuration Yy0 / d1

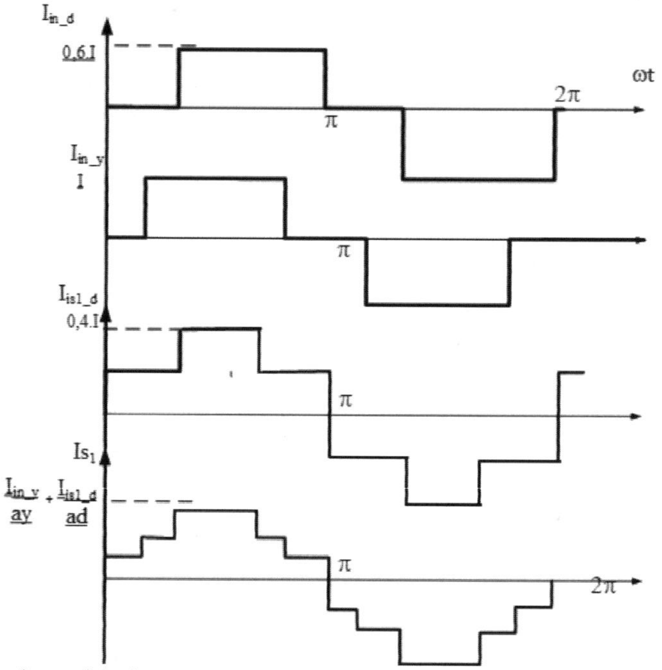

Figure 8. The Waveform of the input current on a 12-pulse converter with transformer input connection Yy0 / d1

From this waveform, it appears that the input current of the 12-pulse converter with the Dy11/d0 and Yd1/y0 input transformer configuration has the same shape, except that in the Yd1/y0 configuration the input current becomes smaller.

3. Results and Discussion

HARMONIC CURRENT INPUT ON 12 PULSE CONVERTER

a. Dy11 / d0 input transformer configuration

As shown in Figure 8, the input current in each converter is I, thus, in the phase winding connected to Y. The phase current is also equal to I. Meanwhile, in winding d the current in the phase winding will be I / √3. On the primary side, namely in winding D, the current in the secondary y will produce a current I / ay where ay = 400/32 = 12.5, while the current in the secondary d will produce a current I / ad where ad = 400/55 = 7, 27. Thus, the phase current of the primary winding D will be equal to $\left(\frac{I}{12{,}5}+\frac{I/\sqrt{3}}{7{,}27}\right)$. The current on the input side of winding D will be equal to $\left(\frac{I}{12,5}+\frac{I/\sqrt{3}}{7,27}\right)\sqrt{3}$ as shown in figure 8 (axis 4). By referring to the figure, the input current in winding D can be stated (Tables 1 and 2).

$$i_s(\omega t) = \sum_{n=1}^{\infty} a_n Sin\ n\omega t + b_n Cos\ n\omega t \qquad (17)$$

And with I = 10 Ampere.

Table 1. I_s (ωt) =

sampai dengan wt = (rad)	0,524	1,048	2,095	2,619	3,143	3,667	4,190	5,238	5,762	6,286
Arus (Ampere)	0,7	2,1	2,8	2,1	0,7	-0,7	-2,1	-2,8	-2,1	-0,7

The solution to this equation is stated in the following table

Table 2. Harmonic order

Harmonic Order	1	3	5	7	9	11	13	15	17	19
c	2,9	0,0	0,0	0,0	0,0	0,3	0,2	0,0	0,0	0,0
a	2,88	0,00	-0,04	-0,03	0,00	0,26	0,22	0,00	-0,01	-0,01
b	0,00	0,00	0,00	0,00	0,00	0,00	0,00	0,00	0,00	0,00
theta	-90	-90	-90	-89	-89	-89	-89	-89	-89	-89

Thus, the converter input current equation at the Dy11 / d0 transformer connection can be stated, $i_{in\_d}(\omega t) = \sum_{n=1,3,5}^{\infty} c_n \sin n(\omega t + \theta)$. Harmonic values of the order 3, 5, 7, and 9 are below 2% so that they can be ignored, then the 12 pulse converter input current can be stated:

$$i_{in_d}(\omega t) = 2,9\ \sin(\omega t - 90) + 0,3 \sin\ 11(\omega t - 90) + 0,2 \sin\ 13(\omega t - 90) + 0 + 0 + \text{.....etc.}$$

With the effective current rating $I_{eff} = (2,9^2 + 0,3^2 + 0,2^2)^{1/2} = 2,9$ Ampere, and Total Harmonic Distortion $THD_I = \frac{(0,3^2+0,21^2)^{1/2}}{2,9} = 0,119 = 11,9\%$

The current harmonic spectrum is shown in Figure 9.

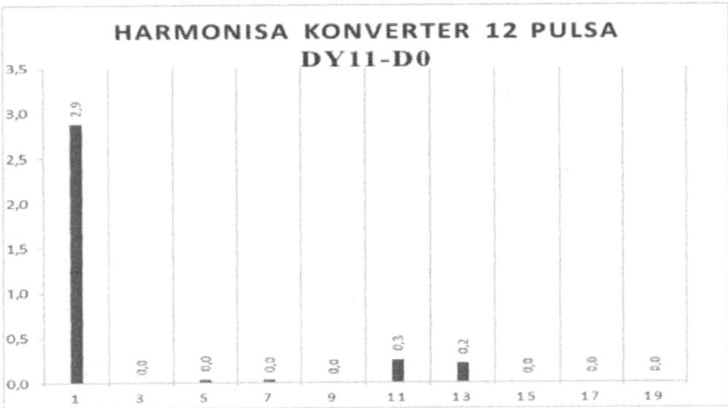

Figure 9. Harmonic spectrum for a 12-pulse converter input current with a Dy11/d0 connection input transformer

Simulation Results

Converter Output Voltage (Figure 10)

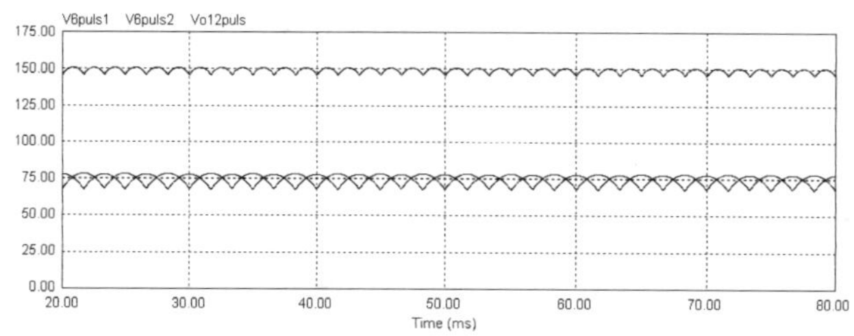

Figure 10. The simulation results of the output voltage waveform converter 12-pulses with the input transformer Dy11/d0 connection.

Input current converter (Figure 11)

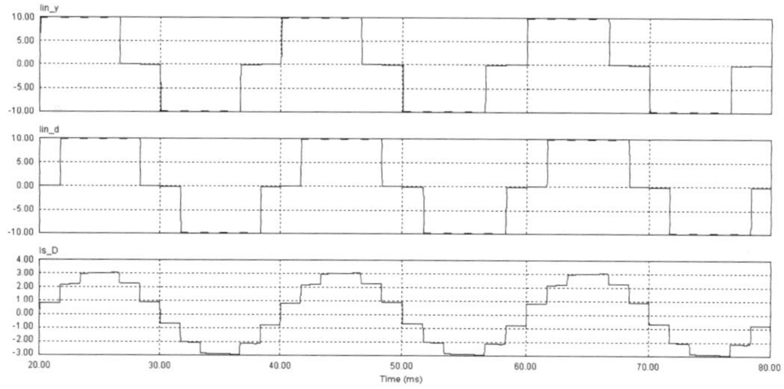

Figure 11. The simulation results of the current 12-pulse converter input waveform with the input transformer Dy11/d0 connection

Input Current Harmonic Spectrum (Figure 12 and Table 3)

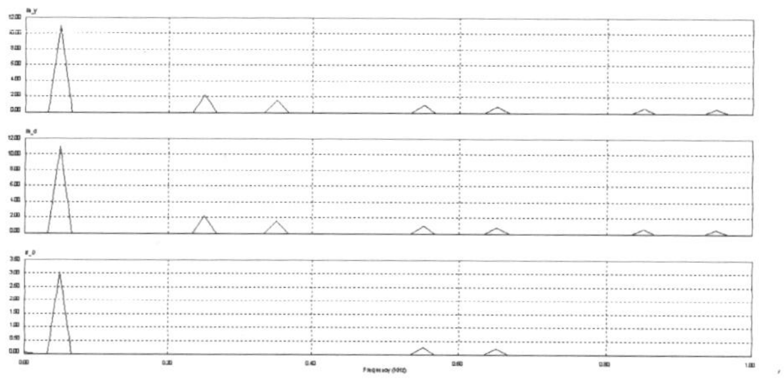

Figure 12. Harmonic spectrum of the 12-pulse converter input current with the input transformer Dy11/d0 connection

Table 3. Simulation result of harmonic order values, effective values and THD

Koneksi Dy11_d0												
Gel.Arus	Fund	h-3	h-5	h-7	h-9	h-11	h-13	h-15	h-17	h-19	Irms	THD (%)
In_y	10,96	0	2,19	1,56	0	0,91	0,79	0	0,64	0,52	11,38	27,92%
In_d	10,96	0	2,19	1,56	0	0,91	0,79	0	0,64	0,52	11,38	27,92%
sumber_D	3,02	0	0,00	0,00	0	0,27	0,21	0	0	0	3,04	11,33%

b. Yd1/y0 input transformer configuration

As shown in Figure 10, the input current in each converter is I / √3. Thus, in the phase winding connected Y, the phase current is also the same as I / √3. Meanwhile in winding d, the current in the phase winding will be equal to I / 3. On the primary side, i.e. in winding Y, it will be equal to $\left(\frac{I/\sqrt{3}}{12{,}5} + \frac{I/3}{7{,}27}\right)$ as shown in figure 10 (axis 4). Referring to this figure, the input current in winding Y can be stated $i_{in\_Y}(\omega t) = \sum_{n=1}^{\infty} a_n Sin\, n\omega t + b_n Cos\, n\omega t$ .... (18)

And with I = 10 amperes (Table 4).

Table 4. $i_{in\_Y}(\omega t) =$

sampai dengan ωt = (rad)	0,524	1,048	2,095	2,619	3,143	3,667	4,190	5,238	5,762	6,286
Arus sumber (30)	0,3	0,7	0,96	0,7	0,3	-0,3	-0,7	-0,96	-0,7	-0,3

The solution of equation (18) is stated in the following table 5

Table 5. The solution of equation

Harmonic Orde	1	3	5	7	9	11	13	15	17	19
c	0,99	0,02	0,02	0,02	0,01	0,10	0,09	0,00	0,01	0,02
a	0,99	0,02	0,02	0,02	0,01	0,09	0,08	0,00	0,01	0,01
b	-0,02	-0,02	0,00	0,00	0,00	-0,04	-0,05	0,00	-0,01	-0,02
Theta	-89,04	-46,98	-84,29	-78,95	-72,13	-64,29	-56,09	-48,17	-41,02	-19,09

Thus, the converter input current equation at the Yd1 transformer connection can be stated,

$$i_{in\_Y}(\omega t) = \sum_{n=1,3,5}^{\infty} c_n \sin n(\omega t + \theta)$$

The values of harmonic currents of the order 3, 5, 7, and 9 are below 2% so that they can be ignored. Then, the 12pulse converter input current can be stated:

$i_{in_Y}(\omega t) = 0{,}99 \sin(\omega t - 90) + 0{,}1 \sin\ 11(\omega t - 64) + 0{,}09 \sin\ 13(\omega t - 56) + 0 + 0 + \ldots\ldots\ldots$ etc.

With effective value

, $I_{eff} = (0{,}99^2 + 0{,}1^2 + 0{,}09^2)^{1/2}$= 1,00 Amp.

With Total Harmonic Distortion, $THD_I = \frac{(0{,}1^2 + 0{,}09^2)^{1/2}}{0{,}99} = 0{,}136 = 13{,}6\%$

The current harmonic spectrum is shown in Figure 13

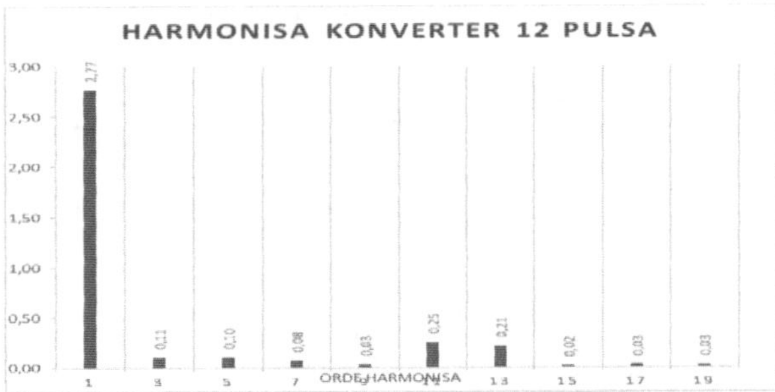

Figure 13. Harmonic spectrum of the 12 pulse converter input current with the input transformer Yd1/y0 connection

Simulation Results

Converter Output Voltage (Figures 14-16 and Table 6)

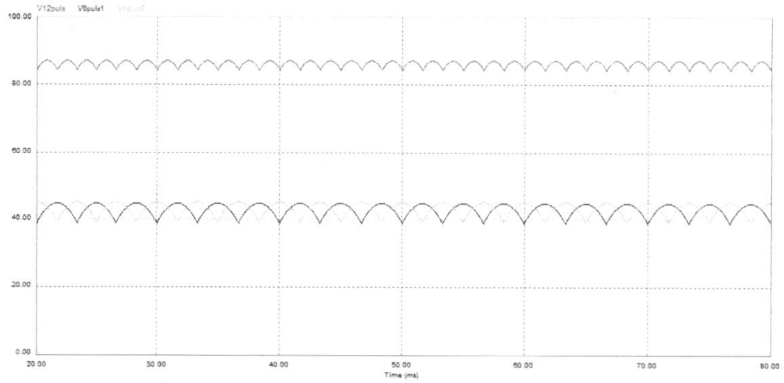

Figure 14. Simulation results of the output voltage wave converter 12-pulses with the input transformer Yd1/y0 connection

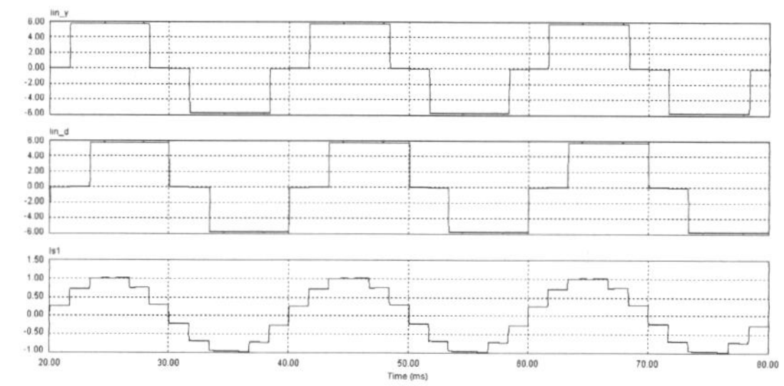

Figure 15. Simulation results of the 12-pulse converter input current waveform with the input transformer Yd1/y0 connection.

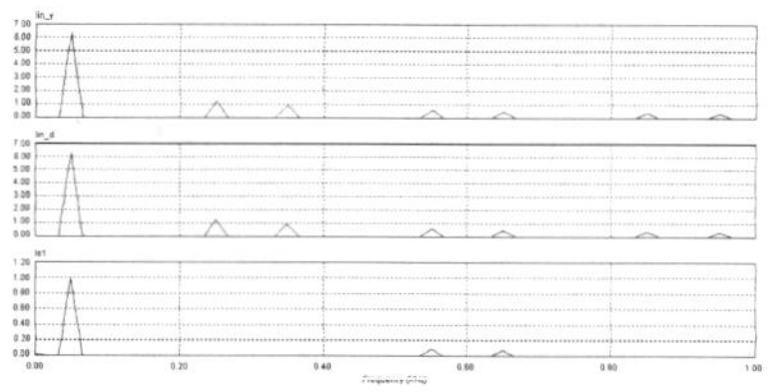

Figure 16. Harmonic spectrum for the 12-pulse converter input current with the input transformer Yd1_y0 connection

Table 6. Harmonic Spectrum

Harmonic Order	1	3	5	7	9	11	13	15	17	19	Irms	THD
c	1,0	0,0	0,0	0,0	0,0	0,1	0,1	0,0	0,0	0,0	1,00	12,03%
a	0,99	0,02	0,02	0,02	0,01	0,09	0,08	0,00	0,01	0,01		
b	0,00	0,00	0,00	0,00	0,00	0,00	0,00	0,00	0,00	0,00		
theta		-90	-90	-89	-89	-89	-89	-89	-89	-89		

4. Conclusion

From the results of the analysis and simulation, it is found that the input side of the two 6 pulse converters obtained thdi of 27.9%, while the 12-pulse converter with Dy11_d0 configuration output current is in accordance with the plan. Besides, on the Yd1_y0 connection the output current becomes 5.8 amperes or 1 / √ 3. Meanwhile, the 12 pulse converter harmonics, in the two input transformer configurations discussed, harmonics of order 3,5,7 and 11 as well as harmonics 15, 17, 19 and 21 can be eliminated, so that only fundamental waves and harmonics of order 11 and 13 alone with the THD value in the two converters relatively the same, ranging from 11.9% to. 13.6%. This shows that the second input transformer configuration does not affect the thdi results, except the power sent by the converter with 12 pulses configuration input transformer Yd1_y0 becomes 1/3 compared to the Dy11/d0 connection

References

[1] M. &. H. (M. Reza Fauzan, ""Analisa Harmonisa akibat pengruh penggunan Converter pada Kereta Rel istrik 1 x 25 KV Jogjakarta – Solo," ELECTRICIAN-*Jurnal Rekayasa dan Teknologi Elektro*, pp. 192 -202.

[2] Ali, M. M., Mochamad, A., and Purnomo, M. H. analysis of Fully Controlled 12 Pulse Converter System and Single Tuned Passive Filter., *Digilab ITS*

[3] Liang, T.-J., Chen, J.-F., Chu, C.-L., and Chen, K.-J. 1999. Analysis of 12 pulse phase control AC/DC converter," in *IEEE 1999 International Conference on Power Electronics and Drive Systems*. PEDS'99 (Cat. No.99TH8475), Hongkong.

[4] Bhim Singh, G. B. V. G. 2006. Harmonic Mitigation Using 12-Pulse AC–DC Converter in Vector-Controlled Induction Motor Drives," *IEEE TRANSACTIONS ON POWER DELIVERY*, Vols. **21**(3), pp. 1483 – 1492.

[5] Paice, D. a. 2000. Transformers formultipulseAC/DC converter, *US. Patent, Rosewood*, Palm Harbor,

[6] H. H. S. 2017. Reduction of Harmonics using 12-pulse AC-DC Converter in Vector controlled Induction Motor Drives, in *International Journal of Innovative Research in Computer and Communication Engineering,* **l**.

[7] Samosir, A. S. 2015. Studi PenggunaanPenyearah18 Pulsa dengan Transformator 3 Fasa menjadi 9 fasa Hubungan segi Enam," *Telokomnika*, **13**, pp. 21 – 32.

[8] Bhim Singh, S. G. 2007. Pulse Multiplication in Autotransformer Based AC-DC Converters," *Journal of Power Electronics*, Vols. **7**(3), pp. 199 – 202\

[9] Singh, B., Vipin Garg and.Bhuvaneswari, d G. 2008. An Input Current Waveshaping AC-DC Converter for Rectifier Loads, *Journal of Power Electronics*, Vols. **8**(1), pp. 1 – 9.

[10] S. C. A. R. v. J. P. N. E. I. Pitel. 1995. Polyphase transformer arrangements with reduced kVA capacities for harmonics current reduction in rectifier type utility interface," in *Proceedings of PESC '95 - Power Electronics Specialist Conference*, Atlanta USA

Bioplastic Technology as Packaging Innovation

Y A Ekawardhani[1]*, C Y Pasaribu[2], A N Rohmah[3], O Salsabila[4]
[1]Departemen Desain Komunikasi Visual, Universitas Komputer Indonesia, Indonesia
[2, 3]Departemen Ilmu Komunikasi, Universitas Komputer Indonesia, Indonesia
[4]Departemen Akuntansi, Universitas Komputer Indonesia, Indonesia

Email: *yully_ambarsih@email.unikom.ac.id

Abstract. This study aims at explaining bioplastics' use as a substitute for plastic base materials. Besides, it is also expected to help sustainable packaging innovation. The use of plastics continues to increase over time. Although plastic is difficult to decompose, its durability makes it still used for various product packaging. Contrary to its use, this resilience becomes an obstacle in protecting the environment from accumulating plastic waste. There are various solutions to replace plastic material with something more environmentally friendly. One of them is the use of bioplastics from cassava peels. This material, called bioplastic, is an innovation to protect the environment in environmentally friendly plastic. This innovation opens up opportunities to be used as a business that connects producers and customers through new technologies and products. The descriptive method used in this research was carried out in two stages: first, in the form of data collection techniques in the form of literature studies through internet resource searches. Second, based on the data obtained, it is used as a reference for determining visual information for packaging made from bioplastic. The visualization process uses Photoshop Creative Cloud 2018 software. The result is an illustration showing the characteristics of the bioplastic material derived from cassava peels and an environmentally friendly plastic material. An illustration in the form of information leads to the formation of conceptual businesses to increase buyer interest and the tendency to use bioplastics.

1. Introduction

The use of plastics as products or packaging for consumer products has become commonplace. Without realizing it, the use of plastic is inseparable from people's habits, both as a product used such as plates or cups and in the form of packaging. Various packaging for everyday products such as powder, lotion, bath soap, laundry soap, or others also uses plastic. Its strong, durable, and lightweight nature is the reason for choosing this material. However, this condition impacts the environment because of its character, which is not easily broken down. The initial idea as a durable material becomes a threat when the amount exceeds what is needed. The idea of replacing plastics has emerged, such as creating a material equivalent to plastic but has fewer negative risks. Bioplastic production is one of the main solutions to reduce problems associated with conventional plastics. They are made from natural materials that can decompose in nature when in contact with microorganisms [1]. Naturally, biodegradable polymers must be broken down into small molecules that are non-toxic and harmful because they are quickly damaged by nature [2]. This bioplastic can be produced in the long term because the basic material is abundant in Indonesia, namely cassava peel. The starch in cassava peels is a natural polymer and additive to modify the starch to make it completely biodegradable [3]. Therefore, starch is considered as one of the main promising ingredients in sustainable bioplastic innovation from cassava peels [4]. Concern for sustainability encourages market demand creation, which unconsciously requires innovation to create new markets [5, 6]. Bioplastic product innovation will attract attention and influence consumer choice and expand consumer reach [7]. The main aspect required is an attractive final product design [8]. Bioplastic innovations include plastic bags, shopping bags, food wrappers, glasses, raincoats in the form of ponchos designed with attractive images.

Bioplastics are considered ideal plastic substitutes for certain types of conventional plastic packaging because petroleum-based plastics leave behind a solid waste material that is difficult to decompose, creating severe global environmental pollution problems. Bioplastics certainly have more advantages than conventional plastics because they come from plants, other biomass, microbial fermentation, or chemical synthesis (Khan 12) [9]. Despite the increasing awareness of sustainability, market demand, and regulations regarding plastics, the development of the bio-plastic packaging industry is still in its early stages. One of the obstacles is the weakness of bioplastics' performance when applied to various kinds of packaging. For example, certain bioplastic packaging's barrier properties, which are very important to protect food products from the outside environment, are often below conventional plastics (Benetto 1) [10]. Bioplastics can also be classified similarly to conventional plastics based on material properties such as in terms of strength and toughness. Based on this, plastic materials can be divided as thermosetting (meaning the cap material is hard and durable) or thermoplastic (compared to thermosetting materials, the material is less rigid) [11]. This study described how innovations in the bioplastics business based on cassava peel starch can play a central role in connecting producers and customers through the development of new technologies and products.

To be a business that produces promising environmentally friendly products, attractive packaging innovation is needed. The purpose of this study is to explain that using bioplastics as a population of plastics is expected to help sustainable embedding and increasingly sophisticated packaging and packaging innovations that will generate public interest in choosing bioplastics. The method used in this research is descriptive method with data analysis techniques, namely literature study and internet search. The method used in this innovative bioplastic packaging design process uses the Photoshop CC 2018 application. Besides, we also design a bioplastic distribution process so that it can be known to the wider community.

2. Method

A descriptive research method is used to get clear results regarding the use of bioplastics as product packaging. The data collection technique is in the form of literature study through the internet regarding bioplastic research development. The data obtained is an initial stage to explain the picture of the development of bioplastics. The second stage is determining the importance of information regarding bioplastics' properties and quality after becoming packaging. Bioplastic materials are similar to plastics, it needs to be marked using visual elements in the form of images. Therefore, people can distinguish between conventional plastics and bioplastics. This stage is carried out by determining an icon that is a sign that the bioplastic material comes from natural fibers and is easily biodegradable. Furthermore, the creation of the image is enhanced using the Photoshop CC 2018 software.

3. Results and Discussion

3.1. Bioplastic Product Innovation

The cultivated bioplastic comes from natural materials, which is an innovative product. Innovation is a form of creativity in developing or creating new products that are valuable and supported by a business model [6]. Innovation, in this regard, is carried out by developing and producing various environmentally friendly bioplastic products and attractive designs. It aims at attracting consumers to use its product. The price of this bioplastic product is somewhat more expensive than conventional plastic. Therefore, there are still many people who choose to use conventional plastic. Many people do not know about the existence of bioplastics. Whereas the use of bioplastics can reduce traditional waste of plastic and improve environmental conditions, various innovative bioplastic products continue to be produced and developed. It is started from plastic bags, shopping bags, food wrappers, fruit and vegetable wrappers, bioplastic cups, to poncho raincoats that are processed from cassava peel starch. In the manufacturing process, plasticizers become additional materials to increase the strength and elasticity of bioplastics. Although this bioplastic product is not much different from ordinary plastic products in terms of shape, we make this product more attractive. So that from plain bioplastic sheets we make innovations so that these bioplastic sheets can be used as products that can help people in their daily lives (See Figures 1 and 2).

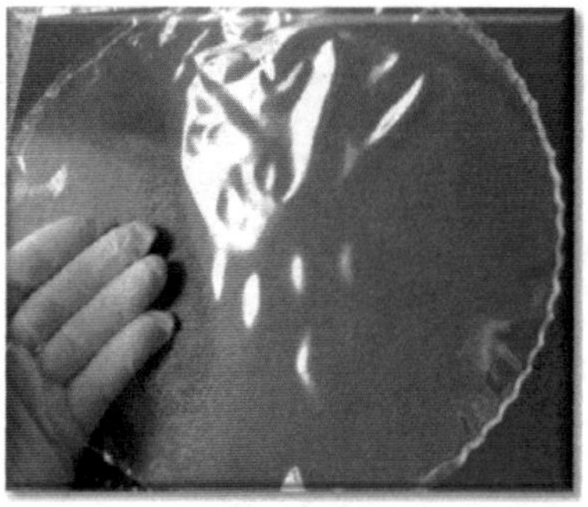

Figure 1. Bioplastic Sheet before Processing [16]

In Figure 2, there are several examples of bioplastic products developed from cassava peel starch in the form of glasses, plastic bags, gloves, and poncho raincoats.

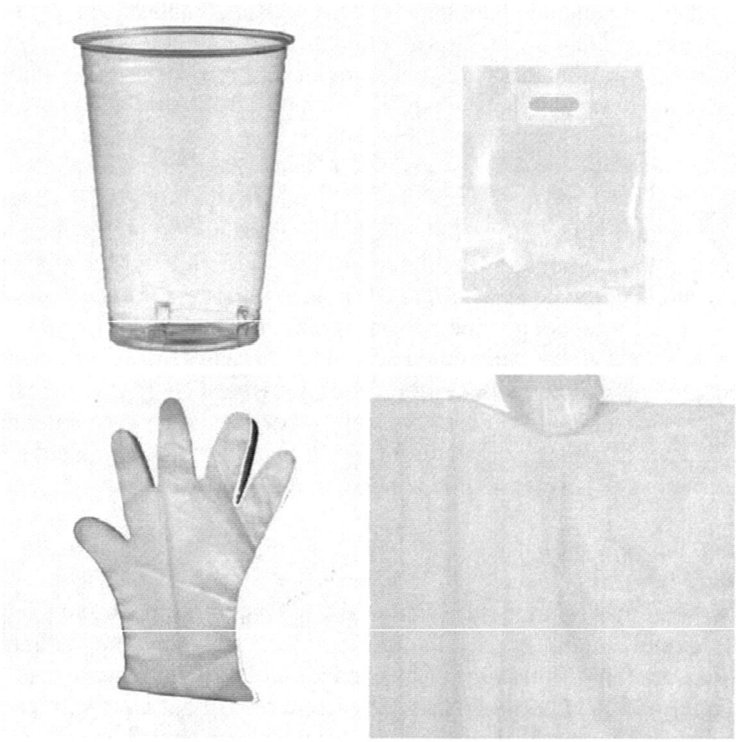

Figure 2. Bioplastic Products After Processing

3.2. Bioplastic Design

We designed a picture that supports environmentally eco-friendly products. The interest in food products with environmentally friendly characteristics and certifications has grown in recent decades. The 'Go green' products that most consumers consider to be low-process, grows na grow, and is halthiare

However, available products that are environmentally friendly are still limited, which makes the use of environmentally eco-friendly products less. Besides, with the development of technology, innovation is also growing. From a plain bioplastic bag, the technology can help to design these bioplastics with attractive designs. Therefore, with the presence of bioplastics with product innovation and attractive designs that describe the raw materials of these products, it is hoped that it will be able to help generate interest from the public to use attractive and environmentally eco-friendly products (See Figure 3).

Figure 3. Bioplastic Product Packaging Logo

The meaning of the logo above is to use environmentally eco-friendly products and products made from environmentally friendly raw materials such as bioplastics which can be innovated into various products not only as plastic bags.

Plastic bags and shopping bags are plastic waste that is often found in many places. It is because many people often use them. Plastic bags are usually used to differentiate in the shop because of their size. It is different from the shopping bag, which can carry more items because of its relatively large size. In accordance with the logo used, this bioplastic bag's design is also designed with an environmentally friendly product atmosphere so that people seeing this product will know that this product is environmentally friendly (See Figure 4).

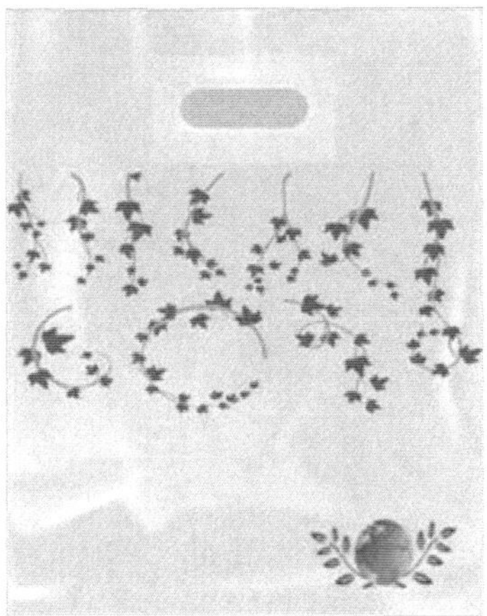

Figure 4. Plastic Bag Design

Plastic cups are also plastic waste, which is widely known because of their disposable use, for example for drinks. This bioplastic glass innovation is intended so that it is used not once or can be stored as decoration because of the attractive product image design. In accordance with the logo used, the design of this bioplastic glass is also designed with an environmentally friendly product atmosphere. When people see this product they will know that this product is an environmentally friendly product (See Figure 5).

Figure 5. Plastic Cup Design

Food, fruit, and vegetable packaging can also be produced from cassava peel starch. The use of bioplastics as packaging for fruits and vegetables can maintain and increase the freshness of fruits and vegetables [13]. A bioplastic food packaging from cassava starch is a food packaging that can be directly consumed. For direct consumption, packaging that is developed is in the form of bioplastic packaging sausages [14].

An innovative raincoat called the bioplastic poncho is also produced, seeing the plastic raincoat waste poncho when the rainy season arrives and many people use it only once. Its relatively large size will be challenging to decompose by nature. The innovation is also in the form of several image designs on the product. Under the logo used, this poncho raincoat design is also designed with an environmentally friendly product atmosphere. When people see this product they will know that this product is an environmentally friendly product (See Figure 6).

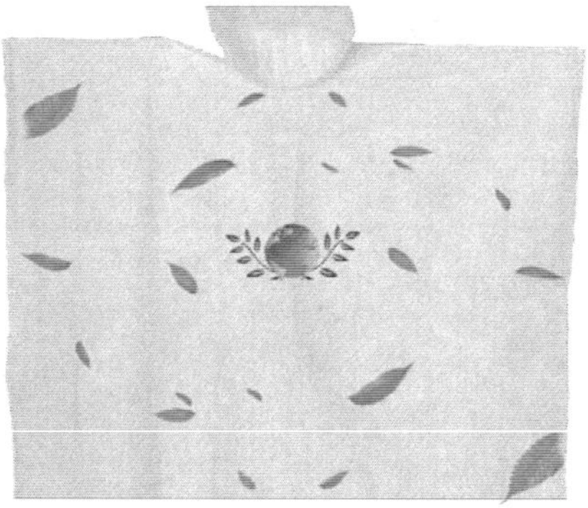

Figure 6. Bioplastic Ponco Design

Gloves are also one of the products that can be produced from cassava starch. During the Covid-19 pandemic, many people used gloves for shopping at supermarkets or carrying out other activities to reduce direct contact with surrounding objects. The innovation here focuses on making people interested in using bioplastic products produced from cassava peel starch with a variety of products and attractive designs, but at affordable prices. In accordance with the logo used, the design of this bioplastic glove is also designed with an environmentally friendly product atmosphere so that when people see this product, they will know that this product is an environmentally friendly product (See Figure 7).

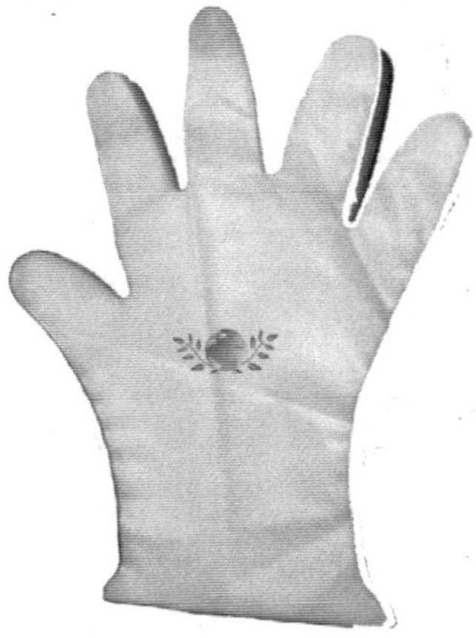

Figure 7. Glove Design

3.3. Product Distribution Design
An innovation that has been introduced and commercialized can generate a lot of development, processes, and technology. When they are developed on a small scale, the emergence of ideas will be different and there will be many problems when they are applied on an industrial scale. Therefore, when starting a product design, it must be adjusted and considered both technology and process for the industrial scale. When the products that have been produced enter the market, the price offered will be slightly higher than those of synthetic plastic bags. Initially, the target market may be aimed more at people who have a back-to-nature concept or industries or companies that prefer biodegradable packaging products. This product's price can continue to decline if technology and process development continue so that the selling price and production price are cheaper. Product design is currently still in the form of a lab-scale, using simple materials and tools to produce a thermoplastic. The product design needs to be tested physically (tensile strength, long time to decompose, etc.). In addition, it is also necessary to carry out an analysis if it is scaled up to the industrial stage, cost analysis, and so on. However, until the laboratory stage, this design is feasible and can be used as a plastic substitute circulating today [15].

The concept of a product affects the level of marketing and consumer purchase value. For that, it is necessary to understand the things that will appeal to the visualization of a product's concept. Visualizing the concept of a product requires visual communication design skills. It means that the design of design needs to pay attention to aspects in making the concept of visualization so that it can be communicative and become a special attraction for consumers. Elements in visual communication design include images, colors, and words (typography). The product concept by introducing new and unusual innovations will certainly affect the purchase value of consumers.

Digital marketing is a process of planning and executing concepts, ideas, prices, promotions, and distribution. In simple terms, it can be interpreted as the development and maintenance of mutually beneficial relationships between consumers and producers. The marketing of products through digital media has many advantages and benefits that can be obtained. Some of the advantages of marketing through digital media include the speed of distribution to consumers, easy evaluation, wide reach, cheap, and effectiveness when compared to conventional marketing to reach consumers. Consumers are everyone who uses or buys a good or service that is within the scope of society for personal gain, other people, or others. In addition, it is not for resale.

A product design to reach consumers must go through several processes. The process begins with product design, concept, and visual design. Then, visual communication, digital media, and reaching consumers is shown in Figure 8.

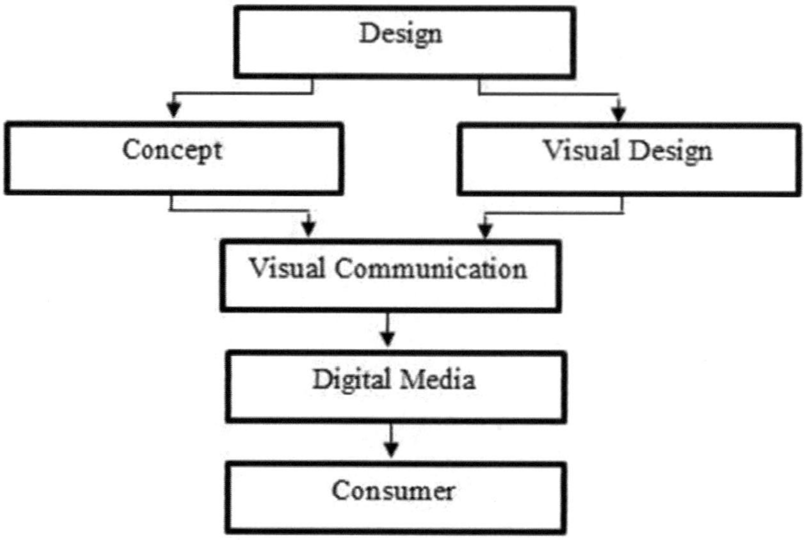

Figure 8. Product Distribution Design

4. Conclusion

Nowadays, the trend of using bioplastics is getting wider. However, it is still difficult for many people to switch from conventional plastics to bioplastics because they are more expensive and difficult to reach. Thus, innovation is needed to get their interest in using products made from bioplastics and with easy access or find. Innovations with designs designed with a contemporary atmosphere are expected to make them interested in buying. The digital marketing provided will make it easier for them to own this bioplastic product.

Acknowledgement

The author would like to thank Ms. Theresia Valentina as a mentor in writing this paper, Universitas Komputer Indonesia who has provided insights and suggestions for this research, and all parties involved in this research, either directly or indirectly.

References

[1] Pulungan, M. H., Kapita, R. A. D., & Dewi, I. A. 2020. Optimisation on the production of biodegradable plastic from starch and cassava peel flour using response surface methodology. *In IOP Conference Series: Earth and Environmental Science*, **475**(1), p. 012019.

[2] Borrowman, C. K., Bücking, M., Göckener, B., Adhikari, R., Saito, K., & Patti, A. F. 2020. LC-MS analysis of the degradation products of a sprayable, biodegradable poly (ester-urethane-urea). *Polymer Degradation and Stability*, pp. 109218.

[3] Tabasum, S., Younas, M., Zaeem, M. A., Majeed, I., Majeed, M., Noreen, A., Zia, K. M. 2019. A review on blending of corn starch with natural and synthetic polymers, and inorganic nanoparticles with mathematical modeling. *International journal of biological macromolecules*, **122**, pp. 969-996.

[4] Ismail, N. A., Mohd Tahir, S., Yahya, N., Abdul Wahid, M. F., Khairuddin, N. E., Hashim, I., ... & Abdullah, M. A. 2016. Synthesis and characterization of biodegradable starch-based bioplastics. *In Materials Science Forum, Trans Tech Publications Ltd*. **846**, pp. 673-678.

[5] Chong, T. Y., Law, M. C., & San Chan, Y. 2020. The Potentials of Corn Waste Lignocellulosic Fibre as an Improved Reinforced Bioplastic Composites. *Journal of Polymers and the Environment*, pp. 1-19.

[6] Geissdoerfer, M., Vladimirova, D., & Evans, S. 2018. Sustainable business model innovation: A review. *Journal of Cleaner Production*, **198**, pp. 401–416.

[7] Xie, X., Huo, J., & Zou, H. 2019. Green process innovation, green product innovation, and corporate financial performance: A content analysis method. *Journal of Business Research*, 101, pp. 697-706.
[8] Shamsuddin, I. M., Jafar, J. A., Shawai, A. S. A., Yusuf, S., Lateefah, M., & Aminu, I. 2017. Bioplastics as better alternative to petroplastics and their role in national sustainability: a review. *Advances in Bioscience and Bioengineering*, **5**(4), pp. 63.
[9] Khan, B., Bilal Khan Niazi, M., Samin, G., Jahan, Z., 2017. Thermoplastic starch: a possible biodegradable food packaging material—a review. *Journal Food Process Engineering*. **40**(3), pp. 1–16.
[10] Benetto, E., Jury, C., Igos, E., Carton, J., Hild, P., Vergne, C., Di Martino, J., 2015. Using atmospheric plasma to design multilayer film from polylactic acid and thermoplastic starch: A screening life cycle assessment. *Journal Cleaner Production*. **87**(1), pp. 953–960.
[11] Ganesh Kumar, A., Anjana, K., Hinduja, M., Sujitha, K., Dharani, G., 2020. Review on plastic wastes in marine environment – biodegradation and biotechnological solutions. Mar. Pollut. Bull. **150**, pp. 110733
[12] Ricci, E, C., Banterle A., Stranieri S. 2018. Trust to Go Green: An Exploration of Consumer Intentions for Eco-friendly Convenience Food. Elsevier Journal. 148, Pp. 54 – 65.
[13] Soegoto, E. S., Jumansyah, R., & Luckyardi, S. (2021). Vetiver Waste Fertilizer As A Growth Catalyst On Mung Bean. *Journal of Engineering Science and Technology*, **16**(1), pp. 165-175.
[14] Saridewi, N., & Malik, M. 2019. Food packaging development of bioplastic from basic waste of cassava peel (manihot uttilisima) and shrimp shell. In IOP Conference Series: Materials Science and Engineering, **602**(1), p. 012053.
[15] Febriani, Elfira. 2017. Inovasi Rancangan Produk dan Bisnis Model Termoplastik Dari Tepung Singkong. *Jurnal Ilmiah Teknik Industri*, **5**(3), pp. 157 – 163.
[16] Araújo, C.S., Rodrigues, A.M.C., Joele, M.R.S.P., Araújo, E.A.F., Lourenço, L.F.H. 2018. Optmizing process parameters to obtain a bioplastic using proteins from fish by products through the response surface methodology. *Elsevier Journal*, pp. 2214-2894.
[17] Kurniati, P. S., Hermawan, Y. A., Rafdhi, A. A., Sabur, A. J., & Saputra, H. 2019. Rain effects on the air pollution (ie, PM1, PM2. 5, PM10, CO, and CO2). *Journal Engineering and Research*, pp. 57-62.

IOP Conf. Series: Materials Science and Engineering 1158 (2021) 012009 doi:10.1088/1757-899X/1158/1/012009

An Android-based COVID Case Monitoring Application Using a Geographic Information System to Educate the Public to be More Aware of COVID-19 Spreads

M C Febriansyah[1], M E Raiz[2], R Zulfikar[3], R D Agustia[4]
Departemen Teknik Informatika, Universitas Komputer Indonesia, Indonesia

Email : chaerrullf@gmail.com

Abstract. The purpose of this study is to monitor the spread of Covid-19 cases. The monitoring process takes advantage of geolocation to detect the level of the spread of Covid in an area. Geographical information system utilization is used for mapping areas that have Covid cases. To notify users of the use of geolocation sensors, it is used to notify system users within a certain radius that the area entered is in an active zone for the spread of Covid-19. To carry out the tracing and tracking process, each Covid-19 patient will be reported to the system by recording their telephone number. Therefore, the patient's movement will be detected whether he complies with the protocol by isolating independently or violating the protocol by exiting the quarantine area. Geolocation sensors and telephone number recording will involve a smartphone device so that the system built will be mobile-based. The software development method used the waterfall method, which consists of requirements, design, implementation, verification, and maintenance. This study is a product design prototype application for monitoring mobile phone-based covid cases using a geographic information system approach.

1. Introduction

WHO has declared Covid-19 as a global pandemic that is harmful to the people. Covid-19 is an infectious disease caused by the newly discovered coronavirus. Coronavirus is part of a large family of viruses called the Coronaviridae family in the Nidovirales order [1]. In mild cases, the virus can cause the infected patient to coughs, experience headache, and flu, while people with previously have known illnesses such as diabetes, cancer, or respiratory disorders can have a higher chance of getting a fatal outcome from this virus infection. The virus was spread through small droplets from the nose or mouth, which is expelled when people sneeze, coughs, or even speak. By seeing how easy it is for the virus to spread [1], the government suggested that people must remain quarantined in their house and maintain a safe distance of 1 meter from each other to interact. Despite the effort of implementing a health protocol, the cases each day still increase significantly. One of the reasons to cause this is there are still many people that disobey the health protocol and have low awareness regarding this virus. The research was conducted to build and develop an android based application called Covitor (Covid Monitoring Case), which has the main focus to educated and raise awareness toward this virus transmission. Covitor will also display the number of confirmed cases and heat map transmission of the virus using Geographic Information System (GIS).

GIS is a framework for gathering, managing, and analyzing data. Rooted in the science of geography, GIS integrates many types of data. It analyzes spatial location and organizes layers of information into visualizations using maps and 3D scenes [2]. This system is usually used in many other areas of expertise, such as making a Malaria vulnerability map or even a symptom tracking [3-4]. Modern GIS technologies use digital information, for which various digitized data creation methods are used. The most common data creation method is digitization, where a hard copy map or survey plan is transferred into a digital medium through the use of a CAD program and geo-referencing capabilities. For the last three months, the application for covid is not many. One of them is web-based and has little publicity [5]. Seeing how many people can have access to Android phones make GIS a great addition to the Covitor app.

Covitor is an application that will give information related to Covid-19. By assuming that most people have an android smartphone, we decide to build our application based on android. Using geographic information systems, the application will have a map that shows the approximation of Covid-19 spread across the area. Covitor will also include a function for the user to enter a custom GPS to point where they are living; once the location is added, Covitor will give a reminder and a warning if the user is going out of their house by tracking the user using the GPS function. The app also has a system that will give warning to the user if the user is going or even enter an area that is heavily populated or has high cases of Covid-19 and finally the app will also educate and commemorate the user surrounding news, symptoms, handling, health policy and raise awareness about Covid-19.

After a few months of practicing the new normal protocol, covid-19 cases seems not lowered much. The spread is still at an alarming and frightening rate. Many people cause this still aren't unaware of the risk of Covid-19 [6]. With that idea, we think of a way to slow down the spread even slightly and maybe eventually stop the growth of Covid-19 cases. With the fast and broad information exchange, use the channel to spread the information on the act on pandemics. A digital platform like social media or Covid-19 related app has proven to be a relevant and effective way to spread awareness and knowledge regarding Covid-19 [7].

Covitor was developed to see how effective it can to educate and raise awareness for people to follow the health protocol given by WHO and the Government. The method that is used in this research is descriptive research, and the model that is used is the waterfall model. The waterfall model was chosen because it creates a linear workflow that fits the development style for the Covitor app. The app is developed using the java programming language in Android Studio. It contains a local database using Room DB to store the user's location and use APIs to collect and display the coronavirus confirm cases, map transmission, and news.

2. Method

This research is focused on developing the Covitor app using the waterfall model, which is one of the software developing methods in the Software Development Life Cycle (SDLC). The waterfall is one of the oldest and most recognized models for building and developing software [8-11]. This software development life cycle progress sequentially. The development cycle can't progress unless the previous step is finished. This guarantees that each process must be done correctly. Generally, the Waterfall model consists of 5 steps can be seen in Figure 1.

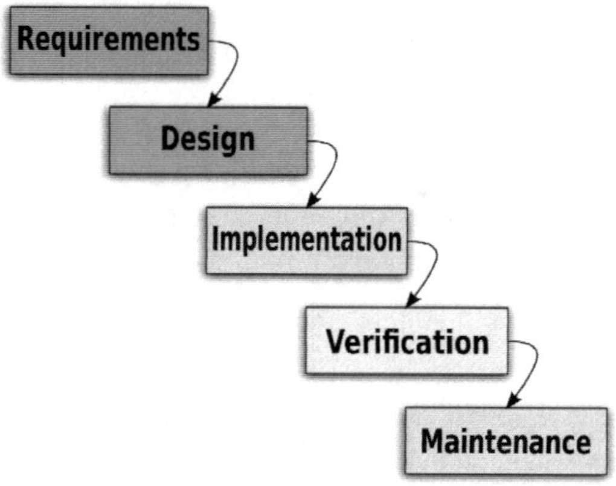

Figure 1. Research Method

1. Requirement and Analysis

In this step, developers must define a description for the system that is going to be developed. These descriptions can come from a client or a certain amount of respondent that are the target users. These descriptions became the requirements that are needed to build the software.

2. Design

The information gathered from analysis and requirement is used to be a guideline and implemented to the general design for the software. It can be a starting point on choosing the functionality, data, data flow, data relation, content, information, and complexity of the software. This phase also a pinpoint where researchers choose the appropriate operating system for the software.

3. Implementation

After getting the general design for the software function, data, procedures, logic, and others, it will be implemented to the desired programing language chosen. In other words, it is building the actual software application itself.

4. Verification and Testing

This phase ensures that software can bring a solution and fulfilled the client or target user needs. It mainly focuses on three big point software and design functionality, as well as navigation. This is why the requirement and analysis process needs to be done correctly.

5. Maintenance and Bug Fixing

Once the product is done, it may still need to fix and adjust. In this phase, the users can express their inconvenience and provide input for improvement. This step also allows the developers to fix some of the bugs and errors in the software.

3. Results and Discussion

3.1. Functionality COVID Monitoring Case System

The specifications for functional software requirements are software requirements resulting from the analysis process carried out in software development. Analysis of software requirements specifications is needed to make use of case descriptions. Analysis of functional software requirements can be seen in Table 1.

Table 1. Specification Software Requirement

Code	Specification Requirement
SKPL-F01	User can view case number
SKPL-F02	User can view mapping of Covid-19 transmission
SKPL-F03	User can get news
SKPL- F04	User can get health Protocol
SKPL- F05	User can scan Covid-19 patient in one area
SKPL-NF01	OS Minimum Requirement is 4.4 Kitkat

3.2. Use Case Diagram

Use case diagram of the COVID Monitoring Case System is shown in Figure 2.

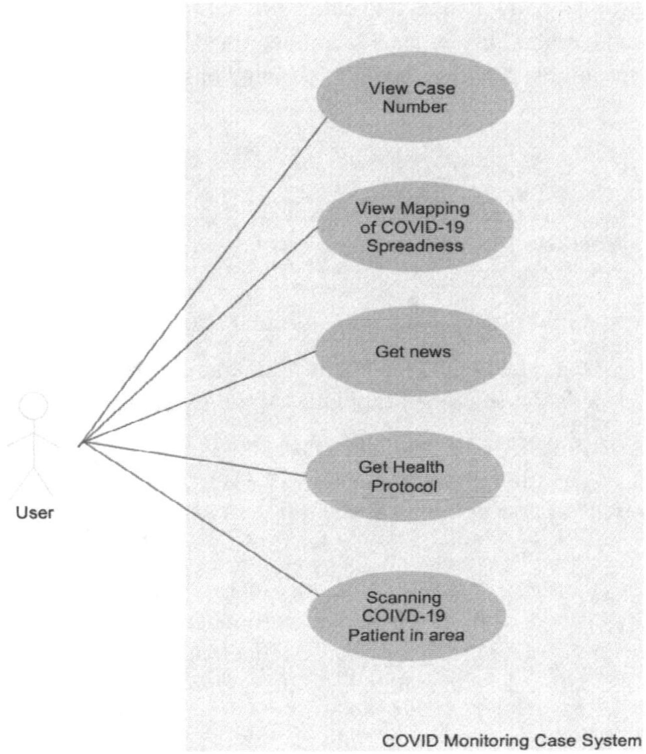

Figure 2. Use Case Diagram

3.3. Implementation
The program code is made in android studio, using Java language. The database that has been using in the program is using SQL. And for the API is using JSON and SQL queries.

3.4. Verification and testing
In this phase, the testing is checking three main aspects:

a) Software functionality
Test whether the functionality in the program is executed well.

b) Program command
Test if the program can run normally.

c) Function logic
Test if the validity of the program is logically true.

3.5. Maintenance and Bug Fixing
After releasing the application, maintenance is still needed to keep the program run normally if some bugs or errors are not shown in the testing.

4. Conclusion
Android phone is one of the most accessible and owned pieces of technology in this day and age. Using it as a platform to spread information and educate people can be very beneficial and effective. In this case, the Covitor app has proven that having an accessible source of news and information regarding Covid-19 can increase the level of awareness and education related to health protocol, Covid-19, and its transmission. Showing and reminding its user about confirm cases, map transmission, and health protocol can minimize the user's desire to go out of their house without protection for unnecessary reasons.

References
[1] Shereen, M. A., Khan, S., Kazmi, A., Bashir, N., & Siddique, R. 2020. COVID-19 infection: Origin, transmission, and characteristics of human coronaviruses. *Journal of advanced research*, ***24***, pp. 91-98.

[2] "what is GIS ?". https://www.esri.com/ (accessed Jul. 28, 2020).

[3] Kim, J. Y., Eun, S. J., & Park, D. K. 2018. Malaria Vulnerability Map Mobile System Development Using GIS-Based Decision-Making Technique. *Mobile Information Systems.*

[4] Yamamoto, K., Takahashi, T., Urasaki, M., Nagayasu, Y., Shimamoto, T., Tateyama, Y., ... & Iwami, T. 2020. Health observation app for COVID-19 symptom tracking integrated with personal health records: proof of concept and practical use study. *JMIR mHealth and uHealth*, ***8***(7), p. e19902.

[5] Boulos, M. N. K., & Geraghty, E. M. 2020. Geographical tracking and mapping of coronavirus disease COVID-19/severe acute respiratory syndrome coronavirus 2 (SARS-CoV-2) epidemic and associated events around the world: how 21st century GIS technologies are supporting the global fight against outbreaks and epidemics. *International Journal Health Geographics.* ***19***(8).

[6] Alzoubi et al, 2020, 'COVID-19 - Knowledge, Attitude and Practice among Medical and Non-Medical University Students in Jordan', *Journal Pure Application Microbiology.* ***14***(1),pp. 17-24

[7] Jena, S., Prajapat, S., & Madaan, R. 2020. Role of Online Platform in Creating Awareness during COVID-19. *Available at SSRN 3637378.*

IOP Conf. Series: Materials Science and Engineering 1158 (2021) 012009 doi:10.1088/1757-899X/1158/1/012009

[8] Alshamrani, A., & Bahattab, A. 2015. A comparison between three SDLC models waterfall model, spiral model, and Incremental/Iterative model. *International Journal of Computer Science Issues (IJCSI)*, ***12***(1), p. 106.

[9] Yuliana, Y. 2020. Corona virus diseases (Covid-19): Sebuah tinjauan literatur. *Wellness And Healthy Magazine*, ***2***(1), pp. 187-192.

[10] Hartini, S., & Dermawan, J. 2017. Implementasi model waterfall pada pengembangan sistem informasi perhitungan nilai mata pelajaran berbasis web pada Sekolah Dasar Al-Azhar Syifa Budi Jatibening. *Paradigma-Jurnal Komputer dan Informatika*, ***19***(2), pp. 142-147.

[11] Anggraeni, S., Maulidina, A., Dewi, M. W., Rahmadianti, S., Rizky, Y. P. C., Arinalhaq, Z. F., Usdiyana, D., Nandiyanto, A.B.D., & Al-Obaidi, A. S. M. 2020. The deployment of drones in sending drugs and patient blood samples COVID-19. *Indonesian Journal of Science and Technology*, ***5***(2), pp. 193-200.

IOP Conf. Series: Materials Science and Engineering 1158 (2021) 012010 doi:10.1088/1757-899X/1158/1/012010

Virtual Reality Utilization as a Character– building of Children in Waste Problems

W S Abioso[1], A I Imam[2*], Y I Maulana[3], M J Prasetyo[4]
[1]Departemen Teknik Arsitektur, Universitas Komputer Indonesia, Indonesia
[2,3]Departemen Desain Komunikasi Visual, Universitas Komputer Indonesia, Indonesia
[4]Departemen Teknik Informatika, Universitas Komputer Indonesia, Indonesia

Email: *alvinakmali@mahasiswa.unikom.ac.id

Abstract. This paper aims to explain the benefits of Virtual Reality (VR) technology as an alternative solution to limited facilities in building children's awareness of waste problems. The method used is a descriptive method that explains how Indonesians treat waste and validates the effectiveness of using VR to develop an awareness of waste. The results showed that children are now more interested in using VR in education to care about the earth's environment. This can happen because it is interactive, but the immersive effect impacts the effectiveness of the learning experience felt by children. The use of Virtual Reality is the main solution with the latest technology in overcoming waste problems in Indonesia by building character since childhood.

1. Introduction

Virtual Reality (VR) is a special technology medium for displaying imaginary images to users by completely imagining virtual environments to experience visuals. Therefore, humans can expand the boundaries of their creation, experience, and learning into anything that can be represented in a virtual environment. In fact, this device has fewer limitations and is closer to the real experience than other devices that have been designed up to now [1]. Students' great interest in the use of this technology is clearly observed, presenting a great intrinsic motivation in its use. It has been verified that virtual reality not only offers great possibilities for teaching at many levels, but is also a relatively unexplored area in the beginning, which is indispensable for deepening its application in the classroom. A new generation of students and learners will naturally learn and interact in a digital environment. In addition, their communicative style requires tools and applications that are adapted to their learning styles [2]. Animation in VR has reasons that will make it more interesting, among others: the resulting story will be simpler and more enjoyable, interacting because it can affect the audience, can reduce costs in video production, can provide a different experience even though watching videos, marketing becomes more interesting [3]. 7 to 11 years old is a golden age where the full potential can be developed and maximally empowered to find something great, innovation, and concepts that can help human civilization between whether a child can do something without adult help and whether a child can do something in adult direction or in collaboration with peers [4]. One of the most important potential uses of VR is found in education. The potential of a learning device that is predicted by a student when he is faced with factual experiences of nearly real learning content can presuppose learning that is much more solid, fast and effective than other traditional learning situations [5].

The purpose of making VR is to invite elementary school children to understand, select and separate inorganic, organic, and flammable waste in an effort to manage and recycle waste to preserve nature through VR technology as an alternative solution to these limitations. Through the Mozilla Hub, along with this activity children will feel motivated because they have friends in one lobby and parents or teachers that can help in character building in children. The method used descriptive method that explains Indonesian society's current behavior and awareness towards waste and validates whether the formation of awareness of waste problems is effective by using VR for children.

2. Method

The methodology used in this research is descriptive method which explains the phenomenon of existing conditions and then explains the literature results. The literature explanation will be adjusted to the results of observations with qualitative interviews as final validation to gain insight (Figure 1).

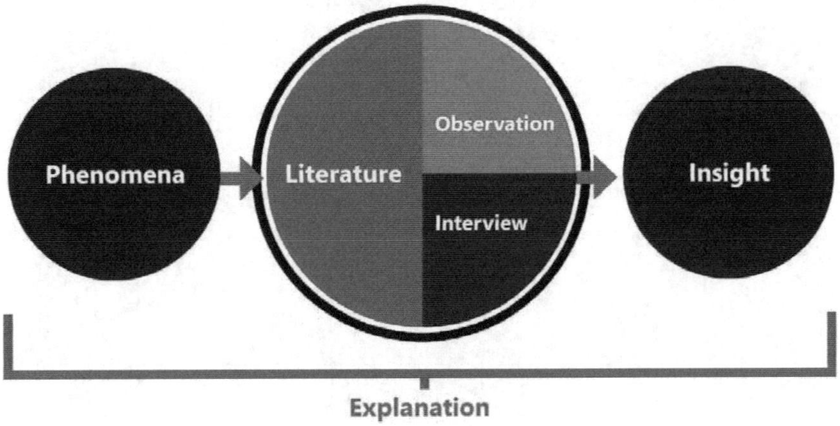

Figure 1. Diagram of Research Methodology
Source: Personal Document (2020)

Observations and research were carried out within 2 months, from June to July in Indonesia particularly West Java (Bandung and Subang). Qualitative data was collected by interviewing the number of six children respondents, who were the objects of research whom generally used to play video games. The method of collecting research objects carried out by direct interviews designed for approximately 10 minutes. Supporting tools for this research are smart phones, internet, and technology from Mozilla Hubs.

3. Results and Discussion

3.1. Indonesian Citizens' Habits towards Waste Management Phenomena

The habit of people throwing garbage everywhere seems to be commonplace in Indonesia. This is exacerbated by the absence of reprimanding each other to people who dump garbage carelessly. Since childhood, this happened to seeing parents who did not immediately teach their children to throw garbage littered. This is of course still very far from the way how the Indonesian people process waste.

Viewed from a survey conducted by the Katadata Insight Center (KIC) of 354 respondents in large cities, including Jakarta, Bandung, Semarang, Yogyakarta, and Surabaya, it revealed that 50.8% of households who do not sort waste have an opinion. That 79% of them argued that they did not want to bother choosing waste based on its type and thought that the waste would be disposed of in a landfill. This survey also illustrates the way households sort their waste from the remaining proportion before, namely 49.2% sorting their waste through waste sorting with 78% sorting into two categories, 18% sorting into three categories and 5% sorting into four categories. But ideally, there should be a separation between organic, inorganic, and hazardous waste. The remaining food, plastic and paper are dominated by household food. For plastic waste, respondents agreed to cover 78%, for leftover food or compost, including pieces of fruit and vegetable skin. Respondents agreed to reach 62%, households are one of the largest waste producers of the total waste in Indonesia [6,7].

In one hour, Indonesia produces 7,300 tons of waste or 175,000 tons per day. Every year plastic production generates about 8% of world oil production or about 12 million barrels of oil or the equivalent of 14 million trees, more than one million plastic bags are used every minute and 50% of plastic bags are only used once and then immediately discarded and from data only 5% is actually recycled according to the "Director General of Solid Waste Management, Garbage and B3 KLHK" (Figure 2).

Figure 2. The Phenomenon of the Waste Problem in Indonesia
Source : https://theaseanpost.com/article/dying-rivers-asean (2020)

The survey concluded that waste management should be done by sorting waste from the house to become a good habit to apply to children. The misunderstanding of *Tempat Pembuangan Sampah Akhir* (TPA stands for Final Processing Place) is also a problem for the environment. TPA is a place to process and return waste to the environment safely for humans and the environment. The community's perception that misinterpreting TPA as TPA makes the waste pile up even more [6]. This was exacerbated by the results of a study by Sustainable Waste Indonesia (SWI) which revealed that 24 percent of waste in Indonesia was still unmanaged. This means that around 15 million tons pollute the ecosystem and environment because they cannot support it per day. Without character awareness and order, this problem will get worse.

3.2. Alpha Generation Literature
The alpha generation is a generation that has known information technology since birth. This generation can take advantage of current technology to develop learning facilities easily. The alpha generation is characterized by being flexible, accustomed to change, capable of dealing with digital technology and a lot of information because this generation is easily immersed in a world connected with technology and information (See Figure 3). The generation is a critical reform of reflection and understanding to become highly progressive and transformative, more original, and develop unique skills [8-10].

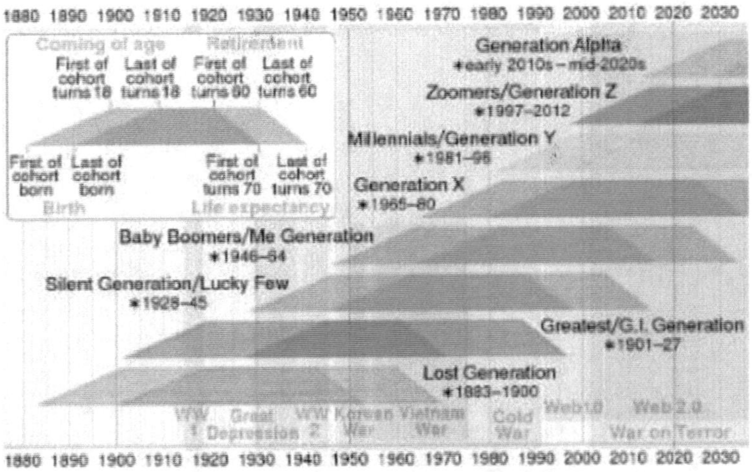

Figure 3. Timeline of Generation Name
Source: https://commons.wikimedia.org/wiki/File:Generation_timeline.svg (2020)

Parents, educators, and other social media have profoundly influenced the concept of connection and have become a central element of this generation. The Alpha generation shows the need for a connection between the interface and technology that switches to the internet, the implementation of touch screen devices for dynamic interaction, therefore the Alpha generation is the controller of the digital world [10,11]. According to Bennet, digital natives are those who have high knowledge and skills in the field of Indonesian technology and this is different from the previous generation. This difference is very pronounced in education, experience and expectations. The Alpha generation prefers audio-visual and animation; the Alpha generation tends to be very addicting to gadgets and technology. The Alpha generation is more adaptable, because they will live in a variety of situations and environments.

3.3. Indonesian Internet Usage

According to Indonesian Internet Service Providers Association (APJII) internet users in Indonesia until year of 2018 have reached 171.17 million. This number increased rapidly by around 27 million users, resulting in 10.2 percent compared to 2017. Java Island became the region with the largest contribution of internet users with 55 percent. Internet usage in Indonesia continues to increase to 184.97 million in 2019. This is predicted to continue to increase if calculated based on statistics sourced from statista.com [4].

According to Newzoo data in 2019, smartphone users in Indonesia have reached more than 82 million and more than 52 million are mobile game players. This puts Indonesia ranked 17th in the world as the most mobile game user and contributed USD 624 million or equivalent to Rp 8.7 trillion for mobile gaming throughout 2019. This proves that children prefer things that have interactions such as games. This game will certainly be closely related to the use of VR (See Figure 4).

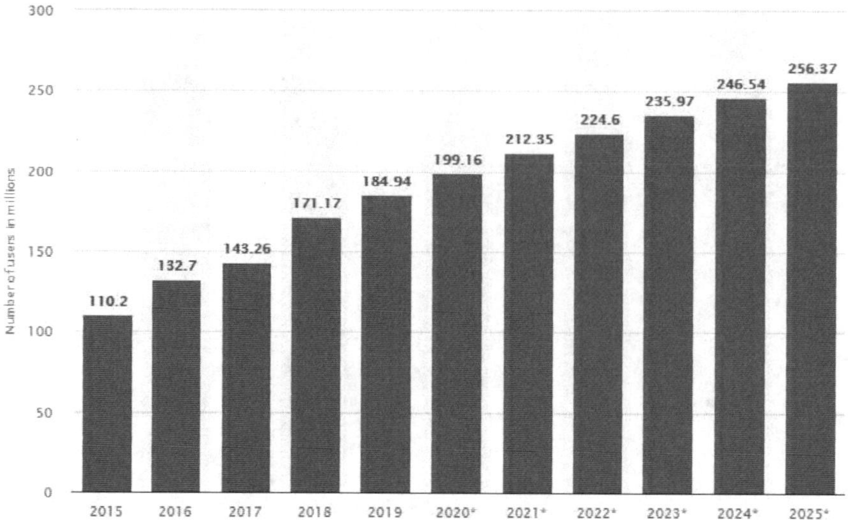

Figure 4. Number of internet users in Indonesia from 2015 to 2025
Source : https://www.statista.com/statistics/254456/number-of-internet-users-in-indonesia (2020)

3.4. Immersive Effect of VR

The immersive effect is the impact of a situation where we feel truly in a virtual environment and wherever you see it is felt around you. VR is a technology that can provide exciting experiences and can lead to different and manifest (physiological) affective (emotional) [8,12]. VR dramatically expands the range of human experience beyond anything that might happen to be encountered in physical reality. VR updating computers is produced by what is seen from a first-person perspective from a third person which is under user control.

There are 2 main features in creating a VR environment: CAVE Automatic Virtual Environment and Head-mounted Display (HMD). A cave is an arrangement that uses a projector and indoor display or similarlarge–scale structure to create an immersive virtual environment. HMD is used on the eye. In HMD, the VR environment will be visible before the user's eyes anywhere when moving (Virtual Reality Society). Research shows that VR can provide added value to learning, especially in terms of motivating students, but the quality of learning is determined by pedagogical science theory and learning that informs virtual environments. His ability makes it possible to learn from experience in various ways, the ability to explore a world that is generally inaccessible, possible, or safe for students. Aligning with social construction approaches to student learning and pedagogy. Its ability makes it possible to learn from experience in a number of ways, the ability to explore a world that is generally inaccessible, possible, or safe for students. VR can present new opportunities for creativity in learning such as role playing and mentoring, opening up learning spaces for practice and exploration, and expanding the ability for inquiry learning. Research shows that the "memory effect" of audio is a contributing factor higher than visual (See Figure 5). Therefore, the application of VR must be accompanied by the delivery of audio for the maximum experience [5][6].

Source	Test	*p*	Cohen's *d*
Effect of Presence on Information Recalled			
Auditory Questions	$t(61) = 2.75$	.008	0.70
Visual Questions	$t(61) = 2.01$	.049	0.51

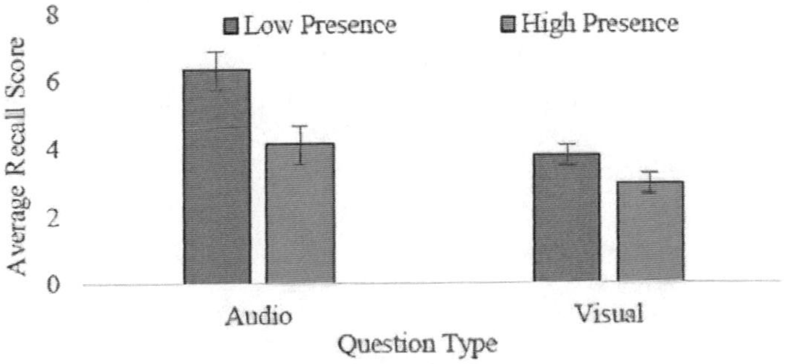

Figure 5. The effects of immersive and future VR expectations on subjective–experiences during an educational 360° video research
Source: https://www.researchgate.net/profile/Daniel_Mcconnell10

3.5. VR validation for Alpha generation in Indonesia

Six Alpha generation children were intervied in Subang and Bandung from June to July for approximately 10 minutes by trying out VR simulations directly. The result shows that more than 80% of them were interested in using VR especially for those already accustomed to playing online games. They feel VR is much more interactive because there is audiovisual to interact with friends through online connection. From different angles, the movement and position of the users of this technology create a feeling of presence or being in a virtual world or in a computer network environment, a shared presence, or a sense of being in a room with other people. To feel being together occurs, there are 2 important factors na: the This strong feeling occurs in a virtual place created. One factor that makes VR interesting since VR has a full immersive impact on users with an unusual experience.

4. Conclusion

Virtual Reality is a technology that can be used for simulations and at the same time can display audiovisuals that interact with everyone in the room or in a virtual world. Since the interest in VR will be something that is attractive so that without realizing it the use of content or learning media will be easily absorbed by users, especially alpha generation children. Using VR as character building can have an impact because of the immersive effect of VR. Suppose children are interested in learning and understanding waste problems through VR. In that case, this will impact character building so that they can solve the problem of waste and its processing in Indonesia. Hopefully it is a one of many solutions that solves the root causes of bad habits in Indonesia.

5. Acknowledgement

The author is thankful to Professor Dr. Ir. H. Eddy Suryanto Soegoto, MT, Rector of Universitas Komputer Indonesia; and also, to Dr. Ir. Herman S Soegoto, MBA, Dean of Faculty of Engineering and Computer Science, for providing significant facilities for the preparation and dissemination of the paper.

References

[1] J. Muller 2020. Indonesia: number of internet users 2015–2025. Retrieved from https://www.statista.com/statistics/254456/number-of-internet-users-in-indonesia/

[2] Gadelha, R. 2018. Revolutionizing Education: The promise of virtual reality. *Childhood Education*, **94**(1), pp.40-43.

[3] Southgate, E., Smith, S. P., & Cheers, H. 2016. Immersed in the future: A roadmap of existing and emerging technology for career exploration.

[4] Ageyev, V. S. 2003. Vygotsky in the mirror of cultural interpretations. *Vygotsky's educational theory in cultural context*, pp.432-449.

[5] Järvinen, A. 2016. Virtual reality as trend contextualising an emerging consumer technology into trend analysis. In *2016 Future Technologies Conference (FTC)* (pp. 1065-1070). IEEE.

[6] Steuer, J. 1992. Defining virtual reality: Dimensions determining telepresence. *Journal of communication*, **42**(4), pp.73-93.

[7] http://documents1.worldbank.org/curated/en/983771527663689822/pdf/Indonesia-Marine-debris-hotspot-rapid-assessment-synthesis-report.pdf

[8] Bahagijo, M. 2019. Indonesia Darurat Sampah: TPA–TPA di Indonesia yang Terancam Penuh. Retrieved from https://waste4change.com/indonesias-waste-emergency-indonesias-landfills-are-on-the-verge-of-overcapacity/2/

[9] Cohen-Hatton, S. R., & Honey, R. C. 2015. Goal-oriented training affects decision-making processes in virtual and simulated fire and rescue environments. *Journal of Experimental Psychology: Applied*, **21**(4), pp.395.

[10] Tootell, H., Freeman, M., & Freeman, A. 2014. Generation alpha at the intersection of technology, play and motivation. In *2014 47th Hawaii International Conference on System Sciences* (pp. 82-90). IEEE.

[11] Wu, W. N., & Kuo, F. Y. B. 2017. Play with google cardboard in a multiplayer environment: how do digital natives and digital immigrants differ?

[12] Suryani, L., & Hamdu, G. Education for Sustainable Development in Science National Exam Questions of Elementary School. *ASEAN Journal of Science and Engineering Education*, **1**(1), pp.1-6.

Designing a Web-based Online Attendance Information System Application

M Iffan[1*], M I Dewantara[2], D Ferdiansyah[3], A J Alexander[4]
[1]Departemen Manajemen Pemasaran, Universitas Komputer Indonesia, Indonesia
[2 3]Departemen Sistem Informasi, Universitas Komputer Indonesia, Indonesia
[4]Departemen Ilmu Komunikasi, Universitas Komputer Indonesia, Indonesia

Email: *m.iffan@email.unikom.ac.id

Abstract. During the current pandemic, many people are working at home or studying online and sometimes have difficulty making absences. Therefore, the purpose of this study is to design a web-based online attendance information system application. The method used in this design was an online questionnaire and interviews to obtain relevant information regarding the attendance list of people who work at home and learn online. The results of this study are an online absence web with several features such as the Global Positioning System (GPS) location of the person who did the absences online, the face of the absent person, and other features. The conclusion obtained in this study is that the existence of web-based online absences can make it easier for people who work at home and learn online to do absences because it is easy and can be done anywhere as long as they have internet access. The benefit of designing a web-based online attendance information system application is the accuracy in providing information to offices and educational institutions. This aims to tackle fraud or falsification of attendance. Online attendance is very much needed and can make it easier for workers and students to do attendance and make it easier for admins to monitor attendance.

1. Introduction

The attendance system is very important in knowing the presence in a company and educational institution. The development of the attendance system currently has developed with the presence of electronics such as computers and gadgets. The attendance system that uses paper has now switched to programs computers, fingerprints, and eye scans that are more practical and accurate. Therefore, the attendance system needs a very good renewal so that students or workers can attend quickly and do not have to spend a long time absent. The best solution is to use a smartphone, computer, and other devices that can access the internet. An attendance system can also maximize time. Therefore, the use of devices that can access the internet is needed in updating the attendance system.

Similar research topics have been carried out by other researchers, such as the Design of Online Attendance Systems Using Android by Husain, which produces online absences that can be accessed on Android [1]. Web-Based Attendance Information Systems by Reymon Rotikan produces web-based online absences. Web-based student attendance application design by Rachmansyah Anessa Putra produces web-based online absences for Mahardhika Junior High School. Web-based employee attendance information system design by Valley Graceynthia Makalare, which produces web-based employee attendance at PT Bank Prisma Dana Manado Headquarters, and also a web-based haul attendance information system by Erliyah Nurul Jannah, who produced a web-based Haul Attendance

Information System at Muhyiddin Islamic Boarding School Surabaya. There were prior researches that described online attendance. Muhammad et al. explain that online attendance applications can help overcome the problem of attendance fraud that often occurs and make it easy to recapitulate attendance [2]. Cloud computing provides functionality for managing information data in a distributed, ubiquitous, and widespread way that supports multiple platforms, systems, and applications [3]. Calimag et al. describe that the use of their mobile gadgets for educational purposes can benefit both students and teachers [4]. Zuru et al. explains that the management of student attendance in the classroom plays an important role in the work of student management. It can help to push students to class on time, increase the efficiency of learning, increase the value of learning, and thus increase the overall level of school education [5]. Sunehra et al. describe that a web page can be used to display the aggregate percentage of attendance of each student according to different classes at the institution [6].

The purpose of this research is to facilitate attendance activities at work or school. In addition, during the COVID-19 pandemic, people must avoid the crowded place. This is easy and practical to use for online absentee assessors. It also helps facilitate data collection for admins because equipped with a feature that can arrange the attendance data.

2. Method

The method used in this research was descriptive research method with the aim to explain and describe the facts systematically, accurately and the results of the research are presented as is. As well as interviews conducted with people around the house and including those at home who experience school and work at home.

3. Results And Discussion

Figure 1 below is a look at the start page of our online timesheet:

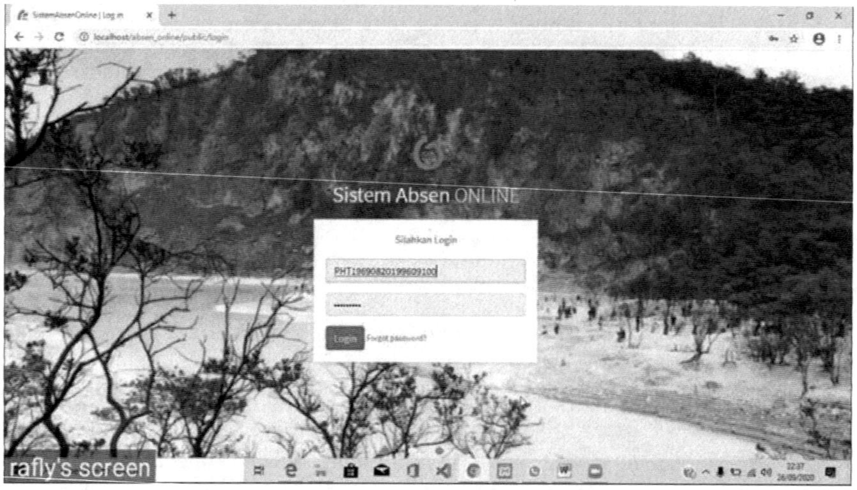

Figure 1. Online Timesheet Start Page

On this page, users can see the initial page for online absences by entering the username and password that have been created by the user. The attendance system provides a convenient way to interact by hosting a web page to view each student's attendance reports [7]. Figure 2 below shows a page when students have done the absence.

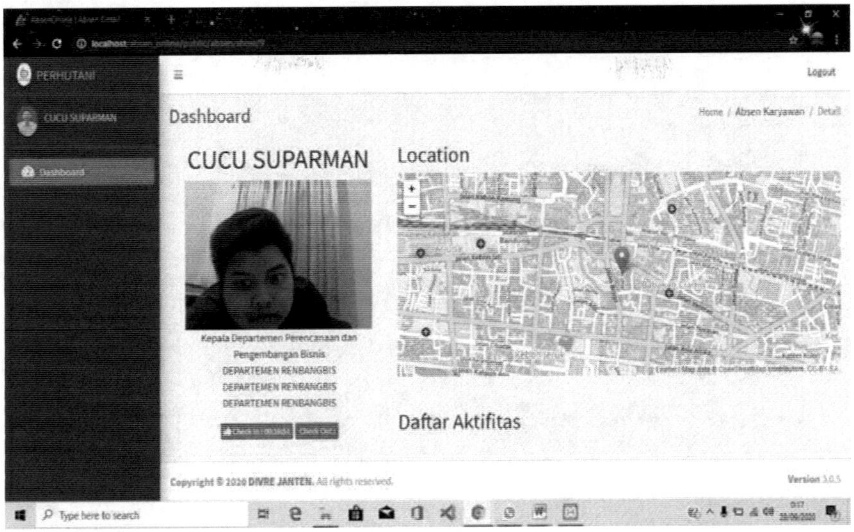

Figure 2. Page When Students Have Done the Absence

On this page, users who have made an absence will get a recap like a picture above, namely a photo of the face, the time of doing absenteeism, and the location where to do the absence. Figure 3 below shows the data recapitulation page.

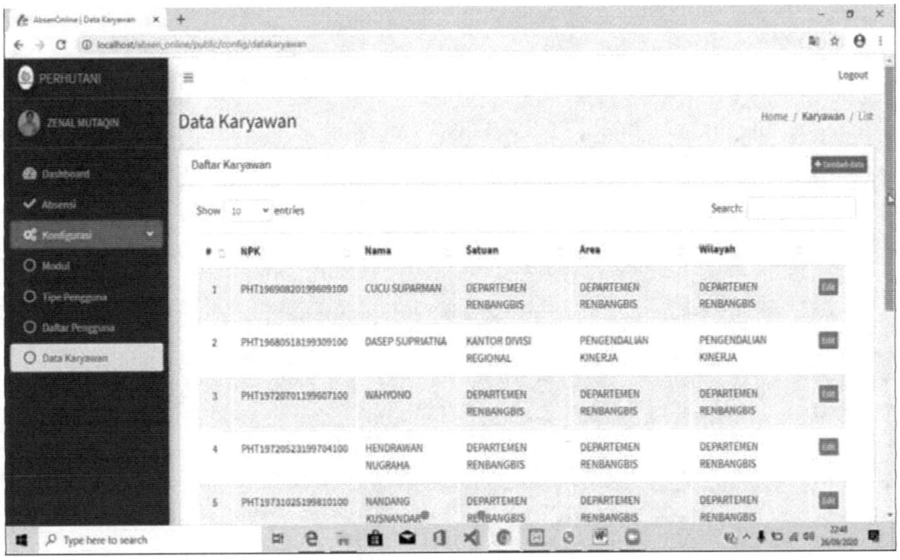

Figure 3. Data Recap Page of Absent Persons

On this page, only admins can access and view the data contained in this section. And on this page, the admin can add, change, and delete profile data on this page. Attendance systems can also be used in applications such as identity management and access control [8]. Figure 4 below shows the flowchart of the online attendance information system.

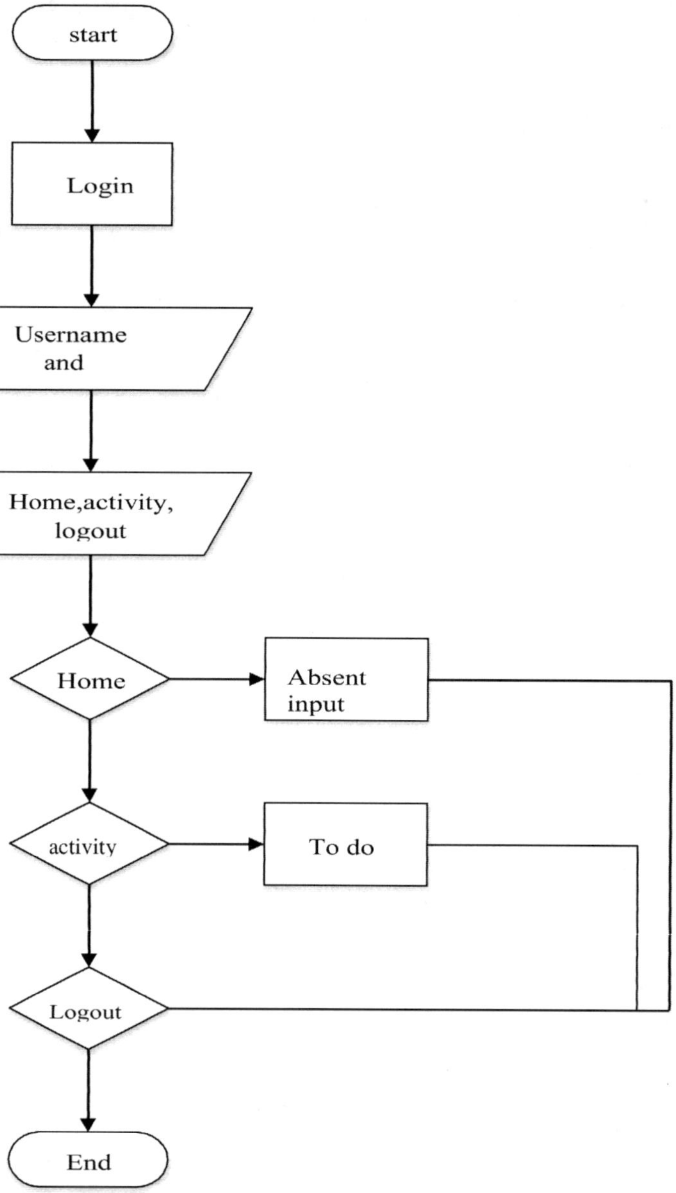

Figure 4. Flowchart of Online Attendance Information System

In an ongoing system, there is a flowchart that is used. This flowchart is related to the procedures/steps for making online absences. An online attendance system can reduce the manpower required for paper-form transmission and shortens document delivery time [9]. Figure 5 shows the activity diagram online timesheet.

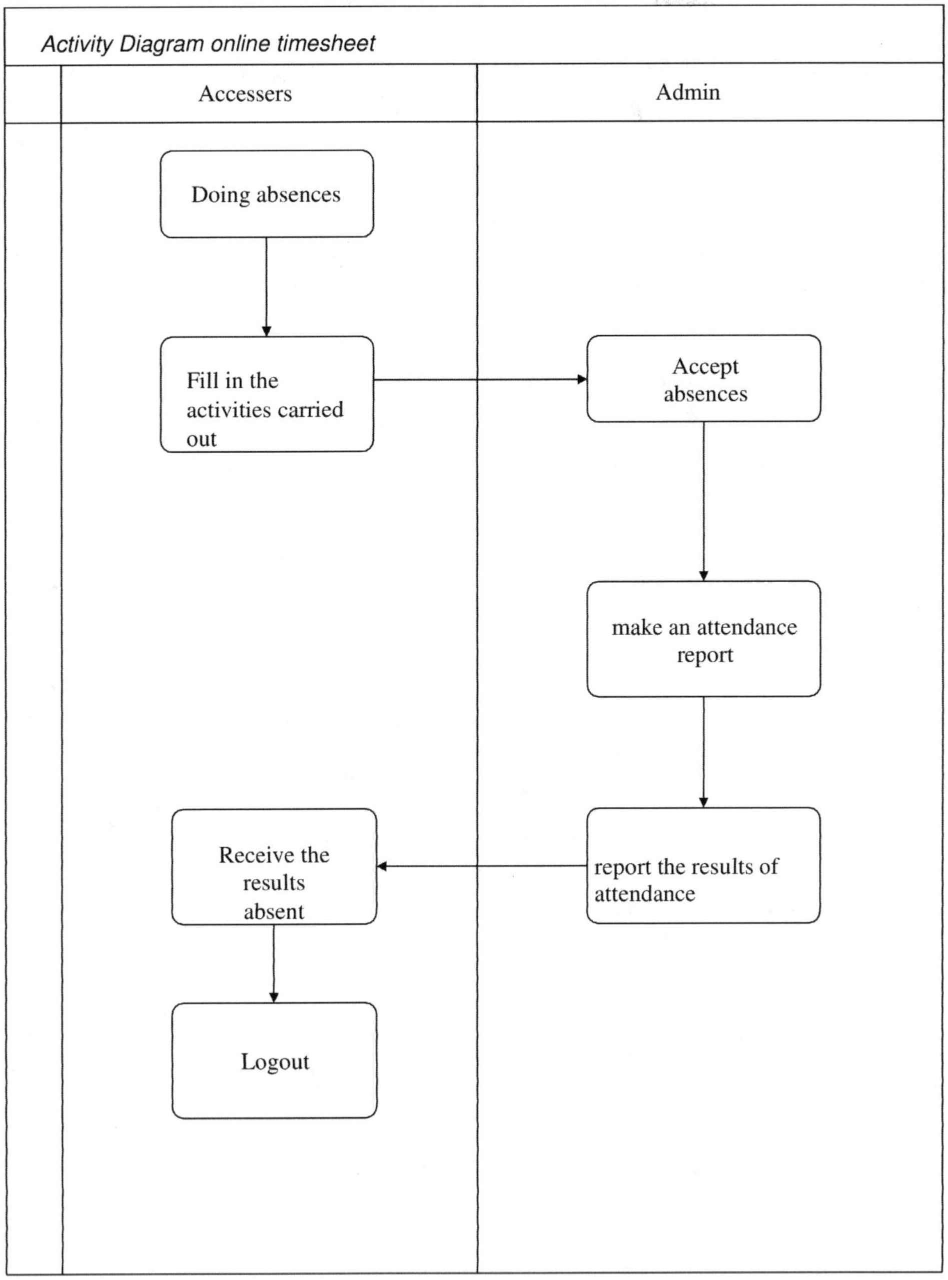

Figure 5. Activity Diagram Online Timesheet

In this session, the researcher described the online attendance flow activity. Above is the activity diagram of the online attendance procedure. The attendance system can keep records of the attendance of students and teachers or staff members of an institution [10].

4. Conclusion

Based on the results of the research and discussion conducted by researchers on the design of online attendance information systems, it can be concluded that online attendance is very much needed and can make it easier for workers and students to do attendance when outside the office / educational institution and make it easier for admins to monitor attendance.

References

[1] Husain, A. Jaka Prima Maulana dan Yuli Widiastuti. 2016. Perancangan Sistem Aplikasi Request Maintenance Teknik Di Perguruan Tinggi Raharja.

[2] Muhammad, N. A., Samopa, F., & Wibowo, R. P. 2013. Pembuatan Aplikasi Presensi Perkuliahan Berbasis Fingerprint (Studi Kasus: Jurusan Sistem Informasi Institut Teknologi Sepuluh Nopember Surabaya). *Jurnal Teknik ITS*, **2**(3), pp. A465-A469.

[3] Doukas, C., Pliakas, T., & Maglogiannis, I. 2010. Mobile healthcare information management utilizing Cloud Computing and Android OS. In 2010 Annual *International Conference of the IEEE Engineering in Medicine and Biology*, pp. 1037-1040.

[4] Calimag, J. N., Mugel, P. A., Conde, R. S., & Aquino, L. B. 2014. Ubquitous learning environment using android mobile application. *International Journal of Research in Engineering & Technology*, **2**(2), pp. 119-128.

[5] Yuru, Z., Delong, C., & Liping, T. 2013. The research and application of college student attendance system based on RFID technology. *International Journal of Control and Automation*, **6**(2), pp. 273-282.

[6] Sunehra, D., & Goud, V. S. 2016, October. Attendance recording and consolidation system using Arduino and Raspberry Pi. *In 2016 International Conference on Signal Processing, Communication, Power and Embedded System (SCOPES)*, pp. 1240-1245.

[7] Kohalli, S. C., Kulkarni, R., Salimath, M., Hegde, M., & Hongal, R. 2016. Smart wireless attendance system. *International Journal of Computer Sciences and Engineering*, **4**(10), pp. 131-137.

[8] Walia, H., & Jain, N. 2016. Fingerprint Based Attendance Systems-A Review. *International Research Journal of Engineering and Technology*, **3**(5), pp. 1166-1171.

[9] Liang, X. Q., Li, W. Y., & Lu, T. C. 2017. A Study of the Multi-Organization Integrated Electronic Attendance System. In *International Conference on Intelligent Information Hiding and Multimedia Signal Processing*, pp. 53-61.

[10] Srinidhi, M. B., & Roy, R. 2015, January. A web-enabled secured system for attendance monitoring and real-time location tracking using Biometric and Radio Frequency Identification (RFID) technology. *In 2015 International Conference on Computer Communication and Informatics (ICCCI)*, pp. 1-5.

Digital Education Program through Accounting Software

U Narimawati[1], J Andriany[2], M Indrianingsih[3], N Octaviany[4*]
[1]Departemen Manajemen, Universitas Komputer Indonesia, Indonesia
[2,3]Departemen Ilmu Komunikasi, Universitas Komputer Indonesia, indonesia
[4]Departemen Akuntansi, Universitas Komputer Indonesia, indonesia

Email: *novyanty.21118097@mahasiswa.unikom.ac.id

Abstract. The purpose of this research is to educate the public about digital MSMEs through Accurate software. We used descriptive methods for 10 MSMEs who have utilized digital platforms but have not utilized Accurate software to support the study. The results show that 4 things affect the facilities and infrastructure of MSMEs' potential guidance process (Medium, Small, and Medium Entrepreneurs) during the practice. From these results, almost 50% of facilities and infrastructure are relevant to MSMEs' skills because they have used the software. MSMEs get experience relevant to digital platforms with Accurate software and affect the potential of facilitating bookkeeping.

1. Introduction

MSMEs have an important role in realizing an independent and fair national economy. MSME's purpose in Indonesia is to realize a balanced national development, justice, and economic structure (Law No. 20 the year 2008 About UMKM) [1]. In business, COVID-19 has made companies instruct workers to work from home [2]. Therefore, MSMEs experience a decline in income has a profound effect on the economy in Indonesia. Indonesia is currently experiencing an economic crisis, so that the Indonesian government is trying to restore its economy with the Digital MSMEs (Micro, Small and Medium Enterprises) program.

Behind MSMEs' significant contribution, there are still many obstacles faced by MSMEs which caused the failure. The low quality of Human Resources (HR), weak marketing, the ability to develop business, limited access to capital, and unconstitutional financial statements contribute to the phenomenon MSMEs failures [1]. Therefore, MSME players need maximum effort to help carry out their books easily through software-based accounting technology. Information technology can meet business information needs timely, quickly, accurately, and relevant [3]. Most MSME players do not understand the importance of accounting information and only see accounting as a bookkeeping process. Nowadays, human information needs are becoming vital. An entrepreneur needs intensive and accurate information for the right decision-making produced by information technology. Business players can use accounting information for decision making. Several factors affect small and medium enterprises in the use of accounting information [4,5]. The use of information technology has significantly affected accounting and financial practices. There have been many changes in manual accounting data to accounting systems that use computers, with accounting software such as Accurate Accounting and other similar programs that can be selected according to the type of business or its business location. Based on Cipta Piranti Sejahtera's (CPSSoft) official website, the Accurate accounting program was first launched in 1999 by CPSSoft. With its vision of "Increasing SMEs in Indonesia into a classy and professionally managed business," the number of Accurate users has been more than 80,000 [6]. The problem of MSMEs in Indonesia is that some ministries, local governments, and other institutions such as state-owned enterprises have provided coaching and assistance. However, the guidance and assistance provided are not on target due to the lack of information on each MSMEs' profile and problems. During this coaching and assistance provided still generalized to all MSMEs, so the results are not as expected [4]. The Digital MSMEs education program is designed to support government programs as the government's efforts to improve Indonesia's economy. Education through the Accurate software system can simplify and speed up financial accounting calculations. To be simpler and more efficient.

The purpose of this research is to educate the public about Digital MSMEs through Accurate software. Socializing the detailed explanation of Digital MSMEs and the use of Accurate software. To support the research, we use descriptive methods to 10 MSMEs who have utilized digital platforms but have not utilized accurate software.

2. Method

This study used descriptive methods with a qualitative approach. Data collection was carried out by observation and in-depth interviews with 10 respondents who are MSMEs in Bandung. The research target is directed at capturing data on identity, characteristics, number, type of business, and SMEs' problems. Sources of information in this study are parties directly involved with MSMEs who have not used Accurate software. Data analysis was carried out using data sources and data types. The interpretations carried out refer to previous studies' results, experts' opinions, and the understanding of the researchers' thinking.

3. Results and Discussion

3.1 Micro and Medium Enterprises (MSMEs)

According to Law No. 20 of 2008, Small and Medium Enterprises (SMEs) are small businesses with a net worth of at most IDR 200.000.000 excluding land and buildings where companies. And a stand-alone effort. Furthermore, in this paper, SMEs will be likened to MSMEs. Indonesia, dominated by Micro, small and medium enterprises (MSMEs), needs to pay special attention to this sector because of MSME's considerable contribution to the national economy. There are at least three significant roles of MSMEs in small communities' lives, namely the means of alleviating people from the brink of poverty, means to level the economy of small people, and providing foreign exchange for the country (Prasetyo & Huda, 2019). In 2018 there were 64,194,057 MSMEs in Indonesia (or about 99 percent of the total business units), and the MSME sector has also employed 116,978,631 workers or about 97 percent of the total workforce in the economic sector. In 2018, MSMEs contributed GDP based on prevailing prices of 61.07% nationally (Kemenkop-UKM, 2018). Figure 1 presents the development of MSMEs in Indonesia from 2010 to 2018. MSMEs in Indonesia increase from year to year [7].

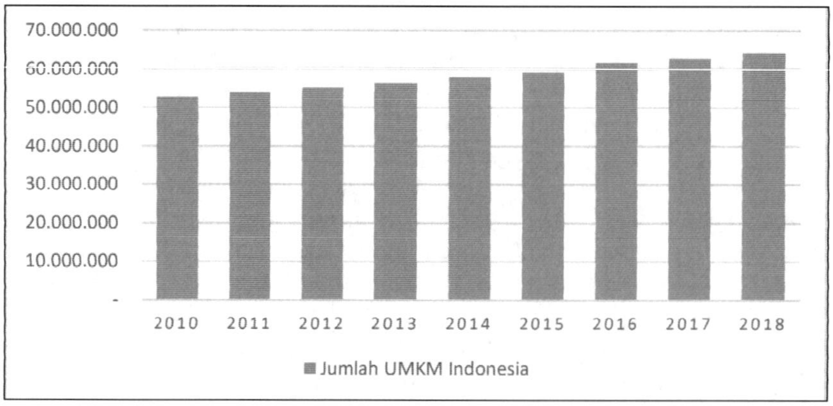

Figure 1. Development of MSMEs in Indonesia in 2010 – 2018

3.2 *MSMEs profile of Bandung city*

Due to the Covid-19 pandemic, around 50 percent of MSMEs in Bandung are estimated to have rolled back. Even so, the government is trying optimally to save MSMEs with various stimuli to suppress the increase in unemployment and poverty levels. At least 40 surveys estimate that half of MSMEs will not be able to survive, the government is trying to revive MSMEs in various ways because there are 60 million MSMEs, not yet the number of workers. The steps taken to restore MSMEs include encouraging MSMEs to accept social assistance, providing tax incentives, relaxation, and loan restructuring, in which 60.6 million MSMEs are already connected with formal financing institutions. Furthermore, providing new loans, including cooperatives, encourages Ministries and Institutions and the development of MSME products and campaigns to buy local products. The government also prioritizes MSMEs' transformation, from relying on offline to online or the digital economy. Today's pandemic requires MSMEs to enter the digital economy. Currently, eight million MSMEs, or 13 percent of the total MSMEs, have been digitally connected. We are targeting that by the end of this year there will be two million MSMEs that can be related to the digital economy, so that there will be a total of 10 million MSMEs. Various surveys show the success rate of MSMEs entering the digital economy is only 4 to 10 percent. There are several, for example, in the online market, MSMEs have to deal with big brands, while management capabilities are still low, production capacity and volume are also relatively small. Bakpia Pathok cases can be pelagic, where there are many MSME actors with a limited production volume. This is where there is a need for brand consolidation, and there is also a need for a joint production house to be efficient.

Bandung regional development is the center of regional growth in West Indonesia besides DKI Jakarta. The synergistic role of Bandung has required an increase in public services both locally and regionally National. The completeness of urban infrastructure in the city of Bandung has an impact on centralized urban activities. Thus demanding field development trading and industrial businesses, namely trade centers (leather goods, jeans, textile, electronics, t-shirts) and industrial centers (rubber, electronics, ceramics, tofu, *tempe*, dolls, hats, bags, screen printing, ceramics, and patchwork). An increase in transportation facilities the opening of the Cipularang toll road has a logical consequence regarding affirmation. The function of Bandung City as a City of Trade Services, especially during the Covid-19 pandemic, MSME can market their products through digital MSME, or you could say through online.

Information Technology (IT) will significantly help MSMEs, especially in marketing problems in the digital market. A millennial can also accompany MSMEs in packaging technology and product quality. Currently, MSMEs need to be able to survive, even win, is a leader or CEO, or an entrepreneur who is capable of adapting to change, consistent in producing, and innovative in creating products.

The number of MSMEs in Bandung continues to increase from year to year, both in terms of the number of business units, labor, and the amount of investment. Micro, Small, and Medium Enterprises are some of the sectors that play a role in building the Indonesian nation's economy. In times of crisis, The global economy is hit now, hitting the business community at the international and business circles in Indonesia. The MSME sector can become a "safety valve" so that workers do not become idle. As an attempt to increasing accountability, conducting reviews on the Main Performance Indicators, with pay attention to performance achievements, problems, and very strategic issues affect the success of an MSME.

3.3 *The Impact of Pandemic on MSMEs in Indonesia*

Indonesia is one of the countries affected primarily on the economic side. The Covid-2019 pandemic has various impacts on the economy, such as difficulties in finding jobs, difficulty meeting the needs of daily life, no income in meeting daily needs, and many difficulties received from all sectors of the economy fields also feel the impact of Covid-19. As of April 17, 2020, 37,000 MSMEs reported themselves to the Ministry of Cooperatives and SMEs affected by the COVID-19 pandemic. According to the data release, MSMEs' difficulties during the pandemic are divided into four problems. First, there is a decrease in sales due to reduced community activities outside as consumers. Second, the difficulty of capital due to the difficult capital turnover due to the declining sales level. Third, there are barriers to product distribution due to restrictions on product distribution in certain regions. Fourth, raw materials are difficult because MSMEs rely on raw materials from other industrial sectors. Table 1 shows that of the four issues, declining sales' impact becomes the most significant problem felt by MSMEs. The pandemic's impact on SMEs is believed to be greater due to the high level of vulnerability and lack of resilience due to limited human resources, suppliers, and options in overhauling the business model [7] (Table 1).

Table 1. Impact of COVID-19 on MSMEs

Impact	Percentage (%)
Decreased Sales	56,0
Difficulties Staining	22,0
Barriers to distribution of products	15,0
Raw material difficulties	4,0

Source: Setiawan (2020b) and Kemenkop-UKM (2020)

3.4 Government's Efforts to Save MSMEs in Indonesia

Some of the literature is used to explore various directions and policies carried out by the Indonesian government in protecting MSMEs from the impact of the COVID-19 pandemic. The Government of Indonesia needs to make multiple efforts to help the community's economy through various policies. The success of the government's policy depends heavily on the support of business components in the community. The community and government must jointly protect the economy from the impact of Covid-19. The rise of MSMEs after COVID-19 is in desperate need of support and all stakeholders. For example, for the tourism sector, business travel partners, travel agents, hospitality, educational institutions, financial institutions, local communities, insurance, and cooperation with similar industries. Hadi recommends that MSME revitalization policy be carried out by increasing synergy between programs and between government agencies. Expanding the variety of modern promotion efforts of MSME products to the domestic and export markets, implementing credit policies with low-interest rates, and simple processes, and encouraging the improvement of MSME supporting facilities and MSME creativity to be highly competitive [7].

3.5 Solution for surviving from pandemic

Business actors need several actions to survive. Changes and arrangements for financial flows in the company must be made to maintain the business. The following are some suggestions that business actors can apply in dealing with current conditions.

1. Manage Cashflow well. Entrepreneurs have to change the way they do business, from attack mode to survival mode. This effort is needed to maintain long-term cash flow, because given that there is not yet much certainty about when the pandemic will end.
2. New marketing strategies. During a pandemic, people's needs are shifting. They are more likely to put aside basic things like clothes or shoes. To attract public interest, one way is to provide a discount on each product.
3. Make use of banking services. Business actors can take advantage of online banking services to make it easier for entrepreneurs to run their businesses.

During the economic crisis, many MSMEs have experienced a decrease in income, so the government has provided one of the solutions by registering Digital MSMEs. But there are still many MSMEs actors who do not understand digital MSMEs, especially MSMEs actors in remote areas. Education is needed about Digital MSMEs, which also work with Kominfo, Diskominfo, BumDes (Badan Usaha MilikDesa) to educate in a structured and comprehensive manner.

The MSMEs program will be guided starting from the online process to online management and promotion. The promotion process itself is listed as a product photoshoot, product description, price, then also gets a domain.id and hosting for free and to make it easier to manage money, it is expected that SMEs will use accurate software. Namely, Indonesian-made, accurate software can help in making financial and tax reports.

The benefits obtained when using accurate are easier to record for bookkeeping, provide print for transaction records that have been made, make financial reports faster, and data security is better accurate with an analysis manager. Accurate software is user-friendly and easy to use because even those who don't have a basic accountant can use it accurately [8-11].

3.6 Advantages and steps using accurate software

The advantages of using Accurate:

1. Save time and speed up work, for example, in entering the sales invoice, it will automatically enter into the sales list, stock items out, and also the loss report because accurate software has simplified the whole system.
2. The working process is easier because there is no need to collect and make formulas manually.
3. Data is safer because it is equipped with data backup.

The steps using accurate software:

1. Install the Accurate 5
2. Create a database
   At this stage, the user can input company data, the currency used, other accounts & lists, customers & suppliers, inventories, and fixed assets.
3. Inputting transactions
   After the end of the month process, MSMEs players can input the transaction. Transactions that are input can be automatically entered in the general journal, not only in public journals. Transactions are also automatically inputted in financial position reports, income statements, cash flow statements, equity statements, trial balances.
   - Purchasing module, consisting of a purchase requisition module, a purchase order module, a receive item module, a purchase invoice module, a purchase item module (purchase payment), and a purchase returns module. So users can easily record the transactions that occur.
   - Sales module. It consists of a sales quotation module, a sales order module, a delivery order module, a sales invoice module, a customer receipt module (sales acceptance), a sales return module.
   - Inventory adjustment is a form containing the activities of adjusting the amount of inventory (inventory) between records in Accurate with the company's manual records (physical quantities available in the warehouse).
   - Journal voucher (public journal evidence) is a form that functions to record transactions that are usually recorded in a general journal where these transactions cannot be recorded using the forms in other modules available in Accurate.
   - Form Payment (payment) is a form that functions to record cash or bank disbursements transactions. Transactions recorded are usually payments for company costs/expenses such as electricity, salaries, telephone, etc.
   - Evidence from this general journal to record transactions, inputting public journal evidence can be found on the General Ledger menu.
   - The information generated, the financial reports generated by the Accurate Software are Profit / Loss, Balance Sheet, Trial Balance, Cash Flows, etc. To see the financial reports generated, first in the explore section click on "Ledger" then select Financial Reports. Select a financial report category then select the desired report details.

4. Conclusion

Based on the above research, it can be concluded that the MSMEs sector in Indonesia is experiencing the effects of the Covid-19 pandemic. This impact can be seen in the economic sector, such as the difficulty in finding jobs. It is challenging to meet everyday life needs, not meeting daily needs, and the many difficulties received from all the economic sectors of Covid-19 all fields. But the government provides economic incentives for MSMEs. Due to the Covid-19 pandemic, many MSMEs are not developing, so education is needed about Digital MSMEs that are also working with Kominfo, Diskominfo, BumDes (*Badan Usaha Milik Desa*) to be able to understand the structure and comprehensively. In addition, to make it easier for MSMEs to manage their finances, it is recommended to use accurate software to make it easier to generate financial reports.

References

[1] Dinanti, A., Wuriasih, D. N., Anggraini, R., Naufalin, L. R., & Nugraha, G. A. 2018. Analysis the Effect of Education on UMKM Performance. *Sustainable Competitive Advantage (SCA)*, **8**(1).
[2] El Junusi, R. 2020. Digital Marketing During the Pandemic Period; A Study of Islamic Perspective. *Journal of Digital Marketing and Halal Industry*, **2**(1), pp.15-28.
[3] Pramiudi, U., & Christiana, V. 2018. The Implementation of Accurate System in Preparation of Financial Statements (Case Study on TB Makmur Jaya). In *THE INTERNATIONAL CONFERENCE ON ACCOUNTING AND MANAGEMENT SCIENCE* (p. 231).
[4] Suryanto, M., & Muhyi, H. A. 2017. Profile and problem of micro, small and medium enterprises in Bandung. In *International Conference on Public Policy, Social Computing and Development 2017 (ICOPOSDev 2017)* (pp. 48-52). Atlantis Press.
[5] Asrida, P. D. 2019. The Impact of Accounting Knowledge and Training towards The Use of Accounting Information on The Owners of MSME in Badung. In *Proceedings of the International Conference on Tourism, Economics, Accounting, Management, and Social Science (TEAMS 2018)*.
[6] Riyadi, M. A., & Rouf, A. 2019. Penerapan Sistem Informasi Accurate Versi 5 terhadap Penyusunan Laporan Keuangan (Studi Kasus Pada CV Percetakan Karya). *Jurnal Ilmiah Akuntansi Kesatuan*, **7**(2), pp.266-277.
[7] Sugiri, D. 2020. Menyelamatkan usaha mikro, kecil dan menengah dari dampak pandemi Covid-19. *Fokus Bisnis: Media Pengkajian Manajemen dan Akuntansi*, **19**(1), pp.76-86.
[8] Putri, R. B. 2020. *ANALISIS PENERAPAN PENGGUNAAN SOFTWARE ACCURATE DALAM PENYUSUNAN LAPORAN PRODUKSI PADA PT. XYZ* (Doctoral dissertation, STIE Mahardhika Surabaya).
[9] Nofie tessar. 2020. KemenkopUKM: 50 Persen UMKM Diperkirakan "Gulung Tikar" Terpapar Pandemi Covid-19. Retrieved February 10, 2021, from: https://m.liputan6.com/news/read/4291018/kemenkopukm-50-persen-umkm-diperkirakan gulung-tikar-terpapar-pandemi-covid-19
[10] Nitisusastro, Mulyadi. 2010. Kewirausahaan & Manajemen Usaha Kecil, Alfabeta, Bandung
[11] Susanta, Gatut dan Syamsuddin. M. Azrin, 2009. Mendirikan & mengelola UMKM, Raih Asa Sukses, Jakarta, Indonesia.

Information Communication and Technology in the Industry

S D Anggadini[1*], M H Ula[2], H Y Darmawan[3], L Yulianti[4]

[1]Departemen Akuntansi, Fakultas Ekonomi dan Bisnis, Universitas Komputer Indonesia, 114-116 Dipatiukur Street, Bandung, Indonesia.
[2,4] Departemen Teknik Arsitektur, Fakultas Teknik dan Ilmu Komputer, Universitas Komputer Indonesia, 114-116 Dipatiukur Street, Bandung, Indonesia.
[3]Departemen Ilmu Komunikasi, Fakultas Ilmu Sosial dan Ilmu Politik, Universitas Komputer Indonesia, 114-116 Dipatiukur Street, Bandung, Indonesia.

Email : *sri.dewi@email.unikom.ac.id

Abstract. This research aims at finding out the role of technology and information in the company. Generally, information and communication technology aims at providing a general understanding of information and communication technology equipment. This research aims to examine the importance of the application of information technology for a government to obtain the value chain of information technology, which is beneficial for all aspects of efforts to improve efficiency and effectiveness. The research method used in this research is a qualitative descriptive method through the socialization of questionnaires and interviews to employees of the Bandung City Government to get more in-depth information. The results prove that information technology has a very important impact on employee performance because it will facilitate all activities to be faster, precise, and accurate, including in the Bandung government. It is because it has utilized modern information technology in managing government management. In conclusion, this research is important to improve the understanding that information technology is necessary and beneficial for the government and win the competition with the private sector. The government must have advantages, core competitiveness, strategies, and innovations, especially in information technology.

1. Introduction

Information and communication technology is a challenge that companies must face. Innovation and applications based on information and communication technology (ICT) have become the main driving force for enhancing organizational performance, economic growth, and social change. [1] Information and communication technologies have become valuable business tools for entrepreneurs [2]. Information and communication technology, education/training, and R&D are among the most important elements for establishing an entrepreneurial platform. [3] Entrepreneurship and ICT require skills to adapt to current economic trends. [4] Sin Tan's Kotelnikov explained the level of entrepreneurs' use of information and communication technology, categorized from basic technologies such as radio and mobile phones to email, e-commerce, and information system processes. Using it is more economically valuable and promising for its business. [5] The company's IT capabilities are essential for better website management because they can portray comprehensive information and electronic payment systems. [6] Multinational companies put ICT business management at risk in terms of post-acquisition integration, and managing the process of shared value creation is the key to success in performance management. Sin Tan also pointed out that the application of ICT in companies can improve business processes through the internet. Entrepreneurs can communicate with customers via email, file sharing, building websites, and e-commerce. [7- 8]

This research refers to previous research. According to Towhich states, information technology impacts various activities, especially when it comes to regulating business processes to adapt information technology to its environment [9]. Like the importance of information technology is applied to the organization[10]. Therefore, it needs an in-depth discussion of what kind of information technology will be applied. Information technology affects internal control because it will manifest more optimal results[11].

Furthermore, activities that include marketing, finance, resources are inseparable from the advancement of information technology because it will facilitate its implementation[12]. Technological advances are impacting its activities to survive against its competitors. This research has differences from previous research, namely the dimensions of content research and revolution problems. Therefore, it is important to be researched [13].

Advances in information technology and communication are bound to affect employee performance. According to Simamora, quoted and translated by Nurhayati, employee performance is the level at which employees meet job requirements [14]. According to Robbins, five indicators measure individual employees' performance, namely quality, quantity, timeliness, effectiveness, and independence. [15] Implementing information technology for a company is to obtain a value chain from information technology, which is beneficial to all aspects of the business and aims at improving efficiency. The research method used in this study is a descriptive method, which is interviews used to obtain more in-depth information.

2. Method

The method used is descriptive and qualitative. The goal is to explain the phenomenon in more depth, proving that the data obtained's detail and depth are very important for research. It is because it uses qualitative research. This research does not focus on the value or number of variables measured. We used interview techniques, observation, and dissemination of questionnaires to the Bandung watershed government in the information system section for data collection. The questionnaire results were collected again to map respondents' answers and compare them with interviews and observations. Then, we analyzed the results to obtain conclusions about the benefits of using information technology in government.

3. Result and Discussion

The use of information and communication technology (ICT) inactivity has grown rapidly in recent years. Nowadays, most people can easily use modern ICT products, namely iPods, mobile phones, personal organizers, and digital TVs to satellite communications technology, personal computers, and the World Wide Web. This new technology can help organizations maximize the search and exchange of information and disseminate it through electronic publications. This allows individuals to directly communicate with others around the world from home or office. [16] It can be shown the useful use of computers and the internet both nearby and far, which is shown in Figure 1.

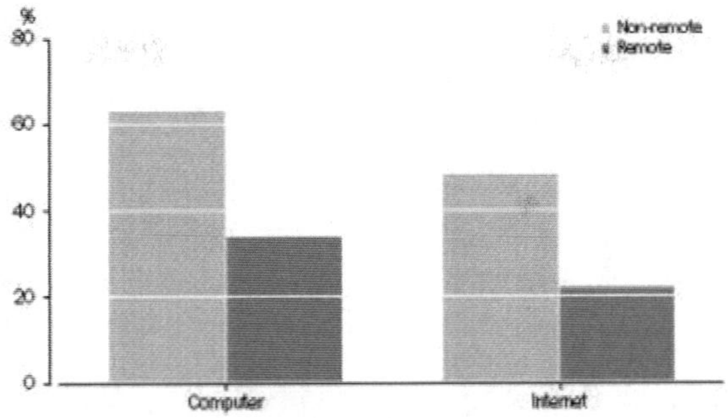

Figure 1. Chart Using of Computers and the Internet in Remote areas

Based on figure 1, it can be seen that information technology is very necessary for both remote and non-long distance because it will help communication and coordination to realize the company's achievements. The revolution in information technology needs to be done so that the company is ready to compete with other companies in its management.

The role of information and communication technology in the field of government is demonstrated by the emergence of e-government, which proven to improve the cooperation relationship of the central government with other parties, such as citizens through Government to Citizen (G2C), business companies through Government to Business Enterprises (G2B), and other governments through government to government (G2G). Good government is a joint agenda that is agreed on to realize a government that costs less, quality government, professional government, digital government, and government that conducts regulatory reforms. Therefore, governments need to build innovation in the public sector, such as developing the policy framework introducing e-government, including the establishment of new strategies and laws involving all stakeholders in the community. The condition of the utilization of information technology in the city of Bandung is categorized as good because it has used modern technology to manage its government [17].

Through this policy, the government responds from various communities; if the public assesses that the policy can improve the quality of administration in Indonesia, then the community's value will increase as well [18]. Anyone can utilize internet-based information and communication technology by knowing the general principle of information search in the internet world. At least two factors determine the amount of time and effort we spend to obtain such information, namely, level of expertise on topics and knowledge of search strategies [19]. A person's level of expertise is very influential in finding information effectively and efficiently as the person can understand the language and prepare knowledge in the field of expertise [20].

Based on the results of surveys and interviews in the Bandung government, the database required by the government of Bandung is by its duties and functions. The naming system of the database used is adjusted to the function of the database in question, for example, the database containing recapitulation data from each work unit and used by the regional head, regional secretary, and secretary of the council for reporting, decision making, and strategic planning of local government named "executive database." To prevent data duplication in some work units, some work units that require the same data must use the same database. Therefore, one database can be used by several work units to build its information system.

4. Conclusion

In the application of information and communication technology, the Bandung Government has been quite complete and has used modern technology. It must be realized that information and communication technology is just a tool that will not create any change if it is not supported by adequate human resources and work culture to run these tools.

Acknowledgment
We would like to thank the parties involved in this research and helped in the distribution of ideas to complete the writing of this research well.

References

[1] Yunis, M., Tarhini, A., & Kassar, A. 2018. The role of ICT and innovation in enhancing organizational performance: The catalysing effect of corporate entrepreneurship. *Journal of Business Research*, **88**, pp. 344-356.
[2] Yuldinawati, L., Tricahyono, D., Anggadwita, G., & Alamanda, D. T. 2018. Towards a framework for ICT-based entrepreneurship development through business incubation processes: case study of a techno park. *International Journal of Business and Globalisation*, **21**(1), pp. 32-45.
[3] Alderete, M. V. 2017. Mobile broadband: A key enabling technology for entrepreneurship? *Journal of Small Business Management*, **55**(2), pp. 254-269.
[4] Atueyi, N. C. 2018. Information communication and entrepreneurship. *Nigerian Journal of Business Education (NIGJBED)*, **4**(1), pp. 294-299.
[5] Ul-Hameed, W., Shabbir, M., Imran, M., Raza, A., & Salman, R. 2019. Remedies of low performance among Pakistani e-logistic companies: The role of firm's IT capability and information communication technology (ICT). *Uncertain Supply Chain Management*, **7**(2), pp. 369-380.
[6] Heim, I., Han, T., & Ghobadian, A. 2018. Value co-creation in ICT services company: a case study of a cross-border acquisition. *Journal of East-West Business*, **24**(4), pp. 319-338.
[7] Tan, K. S., Chong, S. C., Lin, B., & Eze, U. C. 2010. Internet-based ICT adoption among SMEs. *Journal of enterprise information management.*
[8] Basheer, M., Siam, M., Awn, A., & Hassan, S. 2019. Exploring the role of TQM and supply chain practices for firm supply performance in the presence of information technology capabilities and supply chain technology adoption: A case of textile firms in Pakistan. *Uncertain Supply Chain Management*, **7**(2), pp. 275-288.
[9] Janaki, D., Izadbakhsh, H., & Hatefi, S. 2018. The evaluation of supply chain performance in the Oil Products Distribution Company, using information technology indicators and fuzzy TOPSIS technique. *Management Science Letters*, **8**(8), pp. 835-848.
[10] Lee, M., Yun, J. J., Pyka, A., Won, D., Kodama, F., Schiuma, G., ... & Zhao, X. 2018. How to respond to the fourth industrial revolution, or the second information technology revolution? Dynamic new combinations between technology, market, and society through open innovation. *Journal of Open Innovation: Technology, Market, and Complexity*, **4**(3), p. 21.
[11] Mardhatillah, A., & Sari, D. P. 2019. Do Married Woman Employees Need Counseling at Work? A Qualitative Study on the Motivation: Case Study of Sales Women at Information Technology Company. In *International Conference on Psychology and Communication 2018 (ICPC 2018)* (pp. 45-59). Atlantis Press.
[12] Nurhayati, Dina. 2008. Pengaruh Gaya Kepemimpinan dan Iklim Kerja Terhadap Kinerja Karyawan Perusahaan Kerajinan AKP .CRAFT Bantul. Skripsi. Universitas Negeri Yogyakarta.
[13] Robbins, SP 2006. Perilaku Organisasi. Jakarta: Prenhallindo
[14] Petiz, S., Ramos, F., & Roseiro, P. 2015. The use of information and communication technologies in organizational learning practices: A research study in an innovation-oriented Portuguese organization. *International Journal of Advanced Corporate Learning*, **8**(1), p. 4.
[15] Müller, O., Fay, M., & Vom Brocke, J. 2018. The effect of big data and analytics on firm performance: An econometric analysis considering industry characteristics. *Journal of Management Information Systems*, **35**(2), pp. 488-509.

[16] Zhang, J., Li, R., Li, H., Skitmore, M., & Ballesteros-Pérez, P. 2019. Improving the innovation ability of engineering students: a Science and Technology Innovation Community organisation network analysis. *Studies in Higher Education*, pp. 1-15.
[17] Bian, L. 2017. Analysis of Internal Control of TUORI Technology Company. In *2017 7th International Conference on Mechatronics, Computer and Education Informationization (MCEI 2017)* (pp. 218-221). Atlantis Press.
[18] Anggadini, S. D., & Susanto, A. (2017). Study of Resources Allocation on the Implementation of Accounting Information System. *Journal of Engineering and Applied Sciences*, **12**(11), pp. 9255-9259
[19] Akhmetshin, E. M., Vasilev, V. L., Mironov, D. S., Zatsarinnaya, E. I., Romanova, M. V., & Yumashev, A. V. 2018. Internal control system in enterprise management: Analysis and interaction matrices, **2**(3), pp. 102-112.
[20] Al Shobaki, M. J., & Abu-Naser, S. S. 2017. The Role of the Practice of Excellence Strategies in Education to Achieve Sustainable Competitive Advantage to Institutions of Higher Education-Faculty of Engineering and Information Technology at Al-Azhar University in Gaza a Model.

Developing Information System of Population

T Rohmawati[1*], S Khotimah[2*], L Difa[3], B Efrianto[4]
[1] Departemen Ilmu Pemerintahan, Universitas Komputer Indonesia
[2,3]Departemen Akuntansi, Universitas Komputer Indonesia, Indonesia
[4]Departemen Ilmu Komunikasi, Universitas Komputer Indonesia, Indonesia

Email: *tatik.rohmawati@email.unikom.ac.id

Abstract. This research aims to develop information systems in the population administration section in Pahlawan subdistrict in Palembang to facilitate the collection of the population in the subdistrict quickly and accurately. To support the research, we use qualitative methods in descriptions and interviews with several subdistrict employees. The result analysis shows that in Pahlawan subdistrict still uses Information System in Microsoft Excel to record population data; it has not been automated and slow in adding, changing, or deleting data and other population data collections. This study shows that the development of the Visual Studio Information System can provide many conveniences in processing population data. The conclusion of this study is that by creating a population administration information system, the processing of population data such as family card data, birth data, death data, and moving data can be done more quickly and accurately because the data input can be connected well and the data is stored centrally in the database and can facilitate in making population reports based on data that has been inputted into the information system. The impact of this research resulted in a design of an information system that is expected to be used and help Pahlawan subdistrict in recording population data as a useful source of information for decision making in government, especially for Pahlawan subdistrict to be more effective and efficient.

1. Introduction

After the emergence of the phenomenon of globalization and major developments in the field of technology, especially the communication and information revolution, organizations were forced to utilize this technology and use it in developing their business both inside and outside the organization to improve productivity and service operations that must be followed by the organization in the face of competition [1]. Looking at the development of the modern era and the advancement of information technology today, the need for a computerized information system has covered all fields, including the government environment. With increasingly sophisticated technology to create population, information systems are able to have an impact in conducting data processing and are very influential in the efficiency and effectiveness of work [2]. The application of information technology to support daily activities not only pays attention to the needs of the moment but must pay attention to conformity with the vision and mission of the organization as well. This can be seen from some community complaints that the management of population data services is still carried out manually, so that it is considered less effective and efficient [3]. Population Administration Information System is an information system that is compiled based on procedures and uses special standardization aimed at organizing a population administration system to realize administrative order in the field of work [4]. The provision of public services by the apparatus is the embodiment and function of the state apparatus as a public servant in addition to being a servant of the state. In this context, the community is the main actor of development, while the government is obliged to direct, guide, and create an atmosphere that supports the activities of the community. In this condition, the state apparatus is required to be better able to improve its performance and is expected to be better able to formulate concepts or create a conducive climate so that development resources can be a driver of acceleration of the realization of an independent and prosperous society [5].

From surveys in the field showed that the process of public service had used a computerized system of about 30%; the assessment is taken from the data obtained such as population record data either coming or data of the population who moved, the creation of domicile letters, birth data, and death data, where officers must record the applicant's data letter into the led big book and then enter the data into Microsoft Excel which results in the format of the letter becomes capricious and prone to the accuracy of population data recording that has been created and takes a considerable time for each process to occur [2]. From the research of Wijaya et al. it was explained that the implementation of a good information system in the Cimahi City Population Office, it will improve the quality of services at the agency. Population Administration Information System (SIAK) that has been implemented in Cimahi City Population Office has problems when the database is not accessible, so the SIAK application cannot run properly [4]. The research of Alkafi explained that the administration system is very neat, making the office know and understand about the level of service efficiency because it has provided services that cause increased production of goods and services [5]. Innovation is the key to success for the organization to survive and win competition not only in the business sector but also in the public sector, which plays a role and has great potential to contribute to economic growth. Among ASEAN countries, Indonesia ranks sixth behind the Philippines in terms of public services, while Singapore has the highest score of 2.19 [6]. Judging from the benefits provided by SIAK, it is not surprising that the importance of implementing SIAK in Semarang districts' Civil Registry Service Office (Disducapil) properly and correctly. With a population of 996,346 people, it is important for Disdukcapil Semarang Regency to manage an accurate, complete, up-to-date, and integrated population database [7].

Information Technology (IT) is considered an important tool in the constant efforts of public administration to improve efficiency [8]. Basically, the population administration system is part of the state administration system, which has a very important role in the government to manage population data. Everyone is entitled to good and easy population administration services [9]. Therefore, the purpose of this research is to develop a technology in the form of information systems in the population administration section in the Pahlawan subdistrict so that it can facilitate the data collection of the population in the subdistrict quickly and accurately.

2. Method

The method used in this research is the qualitative approach in the form of description and interview to several subdistrict employees. In this research, the administrative information system used is a type of local database. Designing the desktop application is done using Microsoft Visual Studio. Designing the desktop application is done using Microsoft Visual Studio. In Microsoft visual studio, we can use data connected from Microsoft access so that it can connect to Microsoft visual studio easily [10].

3. Results and Discussion

The results of this study were applied to Pahlawan subdistrict in Palembang City. This study used population data as a sample. From the results of the interview with the Secretary of Pahlawan subdistrict, it can be concluded that previously the administrative information system used in the subdistrict was less effective and efficient because the data processing is still using Microsoft excel and there is still a lot of manual data recording so that the processing of population data has not been automated and slow in the process of adding, changing and deleting data and other population data collections. The results of this study show that the development of the Population Administration Information System with Visual Studio can provide many conveniences in the process of processing population data. The results of the discussion also showed that the advantages of administrative information system using visual studio, data input can be well connected and data stored centrally in the database and can facilitate the creation of population reports based on data that has been input into the information system compared to previous information systems that are still manual and use Microsoft excel.

Figure 1 below shows the initial view when opening the visual studio application on the population data administration information system of the Pahlawan subdistrict.

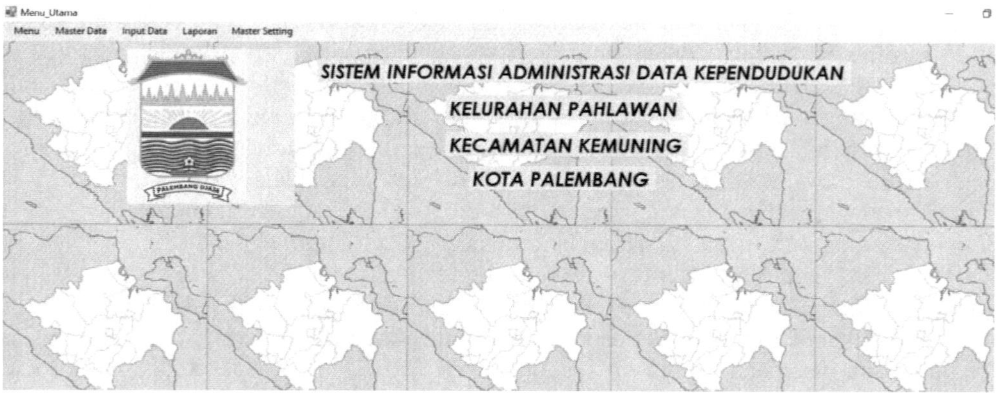

Figure 1. Main Menu Display

The main menu has several sections, namely the menu section, the master data section, the data input section, the report section, and the settings section. The master data section and data input section consists of several data master forms, including population data form, resident identification card data form, family card data form, birth data form, death data form, and population displacement data form. The difference is in the master data section only displays population data that has been stored in the visual studio, while in the data input section, users can input/add new data.

Figure 2 displays the population data form when clicking on the data input section of the main menu.

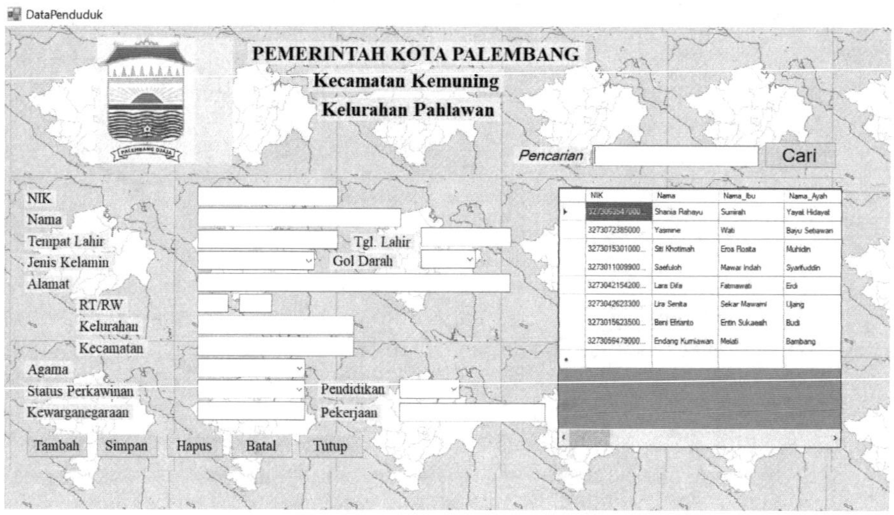

Figure 2. Population Data Display

This form contains the bio of residents who live in the Pahlawan subdistrict. This form is used to search for population data stored in the visual studio by entering the parent number of the population in the search field and pressing the search button. In addition, we can also input new data by filling in the parent number of the residence, name, place of birth, gender, etc., and then pressing the plus button. Figure 3 shows Identity Card Data Display.

Figure 3. Identity Card Data Display

Figure 3 shows a list of identity cards of residents in the Pahlawan subdistrict. Users can input the data of residents who want to create a new resident identification card by filling in the letter number, parent number of the residence, name, address, etc. Figure 4 shows a list of family cards of residents in the Pahlawan subdistrict.

Figure 4. Family Card Data Display

In this page, users can input the data of residents who want to create a new family card by filling in the letter number, letter date, and family card number.

Figure 5. Birth Data Display

Figure 5 shows a list of childbirths for residents who have just given birth. This form is used to record the bio of the newborn child so that the data is stored in the Pahlawan subdistrict. The data that must be inputted is the parent number of the residence, the name of the child, the name of the father and mother, address, name of the helper, etc. See Figure 6.

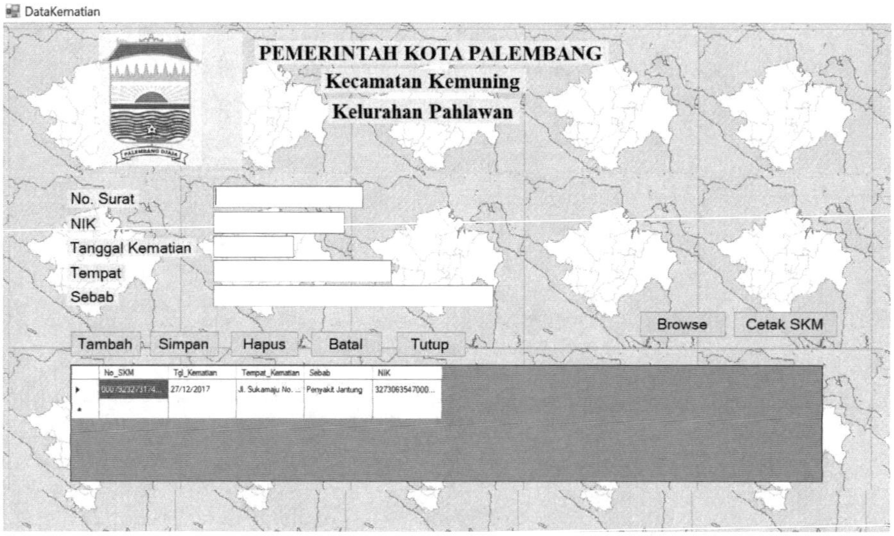

Figure 6. Death Data Display

Figure 6 shows a list of the recently deceased residents. This form is used to record the bio of the deceased population so that the data is stored in the Pahlawan subdistrict. The data that must be inputted is the letter number, the parent number of the residence, the date of death, the place of death, and the cause of death.

Figure 7 below shows a list of population displacements, both mutations from the Pahlawan subdistrict and newcomers.

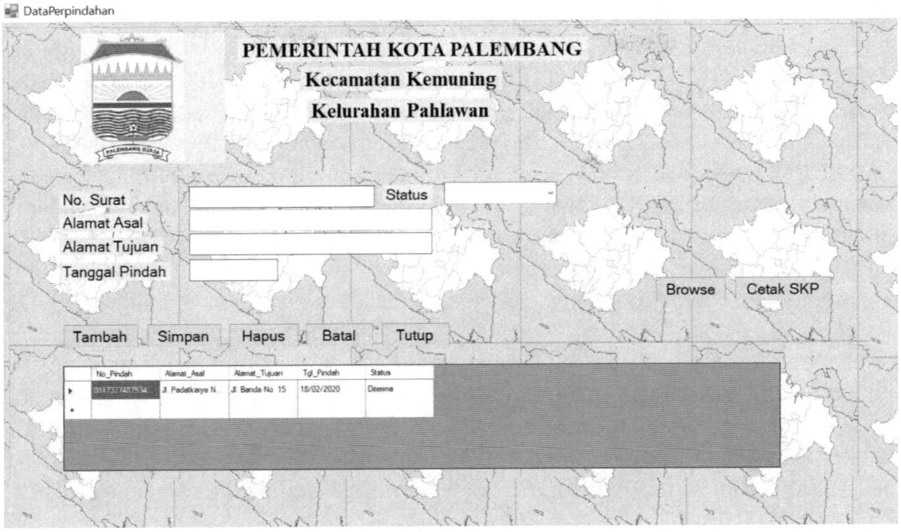

Figure 7. Move Data Display

This form is used to record population displacement data so that the data is stored in the Pahlawan subdistrict. The data that must be inputted are letter number, status, origin address and destination address, and date of transfer.

4. Conclusion

Based on the research, it can be concluded that the creation of a population administration information system can help in the processing of population data such as family card data, birth data, death data, and moving data. The advantages of administrative information system using the visual studio is that the input data can be connected well and the data is stored centrally in the database and can facilitate the creation of population reports based on data that has been inputted into the information system compared to previous information systems that are still manual and use Microsoft excel. With this information, the system can also help to process population data quickly and accurately.

Acknowledgment

We would like to thank the Rector of Universitas Komputer Indonesia for giving us the opportunity to research and write a paper. The author would also like to thank the mentors who have guided in the papermaking process and friends who have helped to complete this paper.

References

[1] Abualloush, S., Bataineh, K., & Aladwan, A. S. 2017. Impact of information systems on innovation (product innovation, process innovation)-field study on the housing bank in Jordon. *International Journal of Business Administration*, **8**(1), pp. 95-105.

[2] Yuli, Y., & Wiyono, W. 2015. Perancangan Sistem Informasi Data Kependudukan Pada Cluster PT. Alam Sutera Realty Tbk. *Jurnal Sisfokom (Sistem Informasi dan Komputer)*, **4**(2), pp. 11-20.

[3] Maulani, Y., & Lestari, S. 2020. Developing Blueprint for Public Services Information System in the District of Indonesia using Enterprise Architecture Planning Method. *JOIV: International Journal on Informatics Visualization*, **4**(4), pp. 178-183.

[4] Istambul, M., & Wijaya, I. 2020. Audit of Population Administration Information System Risk (Case Study: City of Cimahi Population Service). *International Journal of Applied Business and Information Systems*, **4**(1), pp. 86-92.

[5] AlKafi, J. 2020. Kinerja Organisasi Kantor Kelurahan Kemumu Kecamatan Arma Jaya Kabupaten Bengkulu Utara. *International Journal of Policy and Public Administration*, **1**(1), pp. 8-13.
[6] ZA, S. Z., Darma, D. C., Kasuma, J., Ratnasari, S. L., & Tasenţe, T. 2020. Apparatus performance as mediation of creativity and innovation towards the successful application of e-kelurahan. *European Journal of Human Resource Management Studies*, **4**(2), pp. 108-126.
[7] Adriyanto, T., & Kismartini, K. 2016. Evaluasi Kebijakan Sistem Informasi Administrasi Kependudukan (SIAK) di Dinas Kependudukan dan Pencatatan Sipil Kabupaten Semarang. *Journal of Public Policy and Management Review*, **5**(3), pp. 396-409.
[8] Sadiah, H. T. 2020. Website Management Training and ICT Literacy Evaluation of Kelurahan Ciwaringin Bogor Authorities. *International Journal of Quantitative Research and Modeling*, **1**(2), pp. 67-75.
[9] Amalia, E., & Supriatna, Y. 2017. Perancangan Sistem Informasi Administrasi Kependudukan Sebagai Pengembangan Egoverment. *Bandung: Universitas Widyatama*, **2**(1), pp. 81-85
[10] Rubenstein, M., Handaya, D., & Slamet, W. 2018. Design and Implementation of Productivity Monitoring System in Rubber Injection Molding Machine in Real Time Based on Visual Studio and Android Application. In *IOP Conference Series: Materials Science and Engineering*, **384**(1), pp. 012015.

IOP Conf. Series: Materials Science and Engineering 1158 (2021) 012015 doi:10.1088/1757-899X/1158/1/012015

Comparison of Physical and Numerical Model on Dam Break Flood Wave Propagation in Horizontal Channel

V Pratiwi[1*], M S B Kusuma[2], D Kardhana[3], M Farid[4]

[1]Doctoral Program of Civil Engineering, Civil and Environmental Engineering Faculty, Institut Teknologi Bandung, Jl. Ganesha 10 Bandung, Indonesia
[1]Departemen Teknik Sipil, Universitas Komputer Indonesia, Indonesia
[2,3,4]Water Resources Engineering Research Group, Faculty of Civil and Environmental Engineering, Institut Teknologi Bandung, Indonesia
[2,4]Center for Coastal and Marine Development, Institute for Research

Email: *vitta.pratiwi@gmail.com

Abstract. This paper presents experimental results and the development of a numerical model for simulating dam break flood wave propagation. Floods caused by dam breaks have special flow characteristics due to potential energy caused by upstream and downstream elevation differences, and very high flow rates, from the collapse of the dam wall. The physical model is built on a horizontal channel, with 10 meters length and 1-meter width. The reservoirs used are upstream and downstream. Channels used are made of steel materials with negligible friction. The building used is made from wood with a volume of 10x10x40 cm. Flood wave propagation is modelled as a flash flood due to the mass movement of water in large quantities with the sudden opening of water gates. As the initial conditions, the door will be closed. On the upstream of the door, given the initial conditions of water level. Physical models result is used as calibration of numerical models. The developed numeric model is open foam that produces 3D modeling. The result of the physical model and numerical model show the highest flow velocity and good agreement of the pattern of flow velocity profiles during the first 3 seconds. The comparison of the Numerical model and Physical Model indicates whether the numerical model is feasible for use in dam-break modeling. For better results, it is necessary to develop a further model. Solutions to this physical and numerical model can be part of the mitigation effort in case the disaster phenomenon occurs.

1. Introduction

Indonesia is an archipelagic country with a very large population, so the clean water requirement becomes a major problem in the next generation. Therefore, Indonesia needs proper infrastructure to ensure the supply of water needs throughout the year. An alternative water supply infrastructure is a lake. In the rainy season, the lakes will become temporary storage of rainwater to prevent flooding. In the dry season, the lakes will provide water that has been stored during the rainy season. Lake is preferable to the reservoir as it is more constructive in terms of social, economic, and environmental. The advantages of the lake as a water supply infrastructure caused many lakes are built mainly in Jakarta as the capital with a densely populated population. Problems arise when Dam Break occurs in lake that causes a disaster. One of them is the Gintung Lake disaster that occurred in March 2009. Dams can collapse due to various reasons such as insufficient spillway capacity, piping, runoff, and earthquakes [1-2]. The damages caused by the dam failure incidents create disaster in the vicinity area of downstream reach causing loss of human life and loss of properties to the tune of millions to billions [3]. Cases of major floods in residential areas have several issues that need further study. The flow direction that occurs no longer depends entirely on the topographic conditions of the land because of the buildings that block the direction of flow. The flow is direction changed because it hit the building. The modeling of buildings in the case of flooding this settlement is a challenge in flood modeling.

Accordingly, forecasting of severe floods is necessary for prevention and emergency plans, which may avoid live losses and severe damages [4]. The dam break experiment represents idealized conditions, where a gate separating the reservoir and the downstream channel is rapidly removed [5]. The physical model of dam-break flow has mainly involved measurement of temporal and spatial variation of water depth and velocity as the flow propagates as a rarefaction wave [6-12]. Nevertheless, the development of Dam Break's physical model is far from perfect because of the many parameters that affect it. Conceptual approaches are used to minimize the number of parameters that affect. These approaches require sufficient validation in the form of distributed speed water information in the propagation domain. The physical model study is one step to obtain information by doing the running model in the laboratory.

The numerical model dam break was developed to provide an overview of flow conditions as performed on the physical model. So, if the numerical model is feasible and gives results according to the observation, then the numerical model can be used to give the flow forecast due to dam break without the need to do a physical model in the laboratory. Hydraulic models for dam breaking include depth-averaged 2D models and 3D models, both have a mathematical model and numerical model; the other for the dam break model is the clustered model. Some scenarios such as Single Structure Failure Scenarios and Domino Effects (such as the collapse of various hydraulic structures and collapse of buildings) can be assumed [13]. This research is expected to give a result comparison between the physical model and numeric model on channel influenced by flow due to dam break. So, it can be seen whether the numerical model is feasible for use in forecasting flow due to dam break.

2. Method

2.1. Experimental Set-Up and Instrumentation

Dam break experiment facility developed by Kusuma et al. [14] in Civil Engineering Hydraulics Laboratory of Institut Teknologi Bandung. It consists of a rectangular horizontal channel of 10 m length, 0.85 m width, and 0.5 m depth. A reservoir that is used is at upstream and downstream. The channel's base is made of steel which its friction can be ignored. The channel wall is made of flexible glass with a thickness of 8 mm, which visually observes the flow of water. The channel bed is made of frictionless stainless steel. A small rectangular water pool is used as a reservoir in the upper part and as a water recirculation tank in the lower part. The upstream reservoir is made of concrete, 4 m long, 2 m wide, and 0.6 m high. The water recirculation tank is made of steel and is supported by a centrifugal pump and pipe (Figure 1).

Flood flow is modeled as a flash flood caused by the mass movement of large amounts of water with the sudden opening of the sluice gate. As an initial condition, the gate will be closed. In the upstream will give the initial conditions of water height 40 cm. The flow due to dam break will be modelled by a sudden opening of the swing gate (Figure 2). The swing gate made from steel is used to generate a dam-break flow. A dam break flow is generated every time there is a sudden upward swing of the gate. It is recommended to choose a swing gate rather than a vertical liftgate, as the horizontal movement of the swing gate appeared to decrease the failure related to the dam break wave, as rather low failures, were detected with reasonable opening speeds of the gate [15].

Flow velocity was measured using a current meter in Figure 3 supported by a digital data logger connected to the data acquisition system. The velocity measuring instrument used is a current meter with a diameter of 1.5 cm with a high-speed type that can measure the flow at high speed. The velocity measurement begins when the gate is opened until the velocity tool cannot read the value of the velocity of the flow passing at that point. The flow meter is positioned 2 cm from the bottom of the channel to obtain velocity data during low and high flow depth conditions.

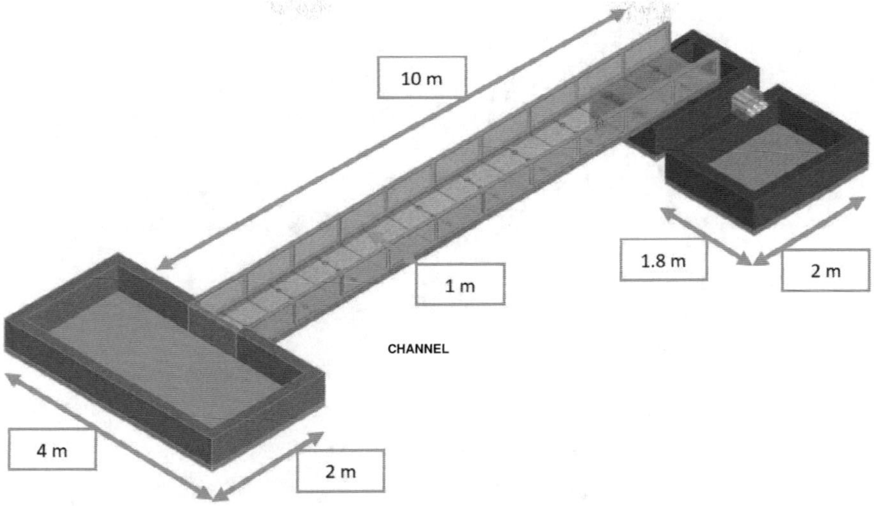

Figure 1. Set up Model

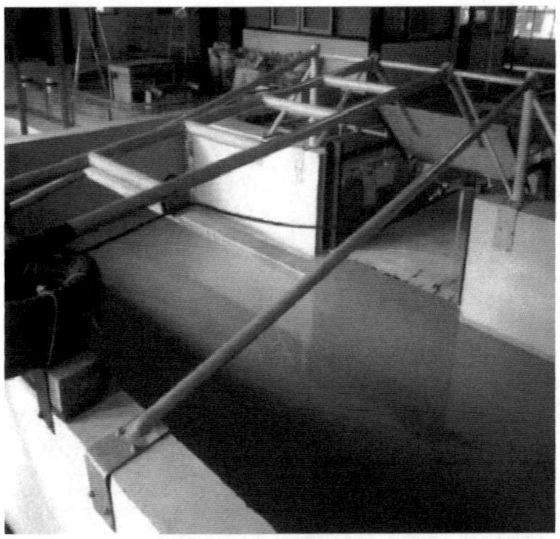

Figure 2. Photograph of Channel and Dam Break Mechanism

Flow velocity meter used is a current meter that uses a fan, a counting machine that calculates the intensity of the rotation in units and the data logger as a recorder of data that records data over time and transfers it in the form of readable data by computer in volts. The current type of fan meter used is high speed (Tool Series No.: B2896), the digital counter used is streamflow-430, Data logger used Data Q (no tool series: DI-710).

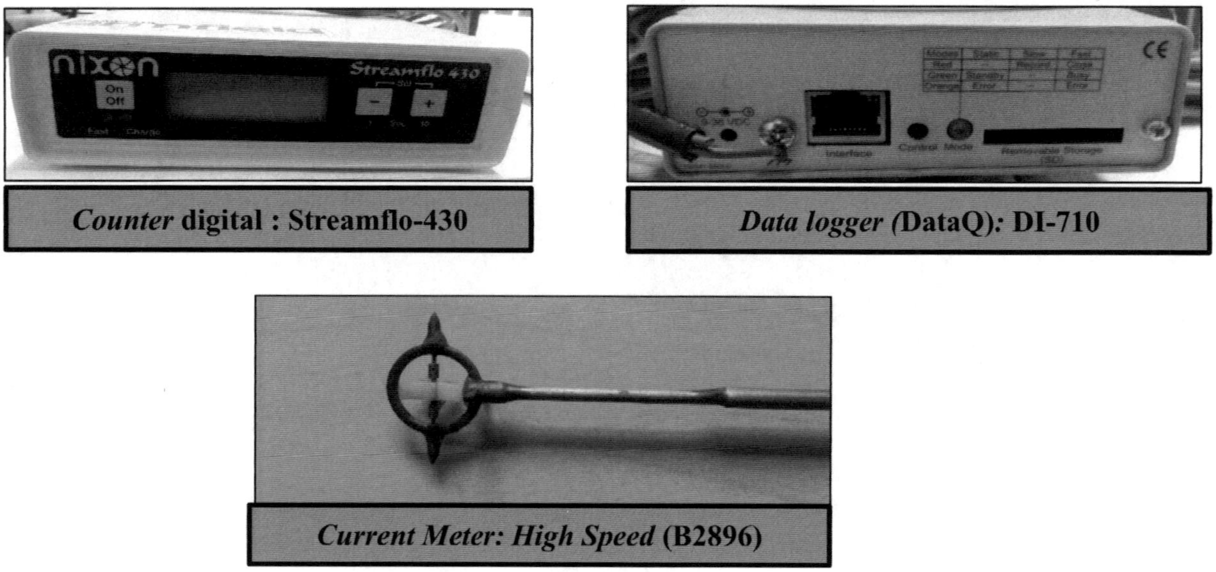

Figure 3. Speed Meter Measurement Tools

2.2. *Scenario Measurement Scenario*
Measurements in the physical model of flood propagation within buildings due to Dam break are done in scenarios without buildings. Seed measurement points are measured from grid 2 to grid 8 and also measurements around the point of the building plan. Point Speed measurement is done at points in Figure 4.

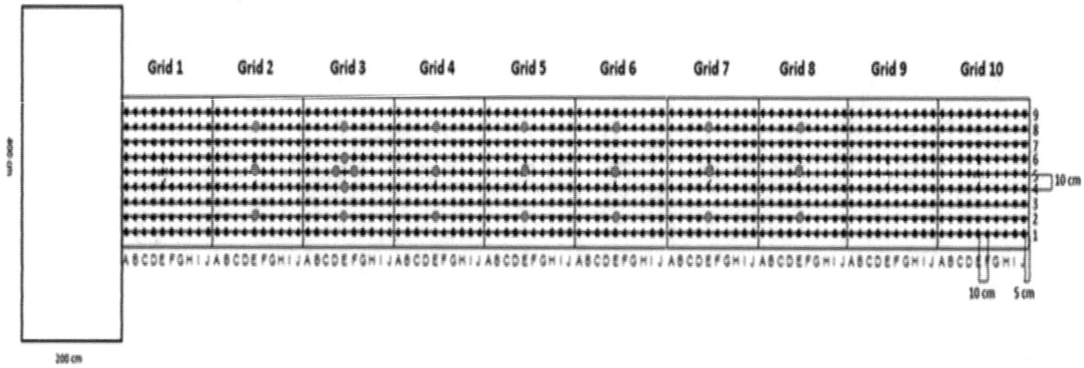

Figure 4. Point of Current Meter Position

2.3. *Numerical Model*
The model used in this study is based on the open-source computational fluid dynamics (CFD) platform called OpenFOAM (OpenCFD, 2008), available for free on the Internet OpenFOAM, primarily designed for problems in continuum mechanics using tonsorial approaches and object-oriented techniques [13]. It is an open-source library (source file in C ++) to solve partial differential equations and enables numerical breakdown adjustments for continuum mechanics with special emphasis on fluid dynamics. To simplify the numeric model and speed up the calculation process, the model is made in half part. This is due to fact in the laboratory, the physical model has a symmetrical shape, so it is considered that the other half has the same condition because it has the same parallel parameters and conditions (See Figure 5).

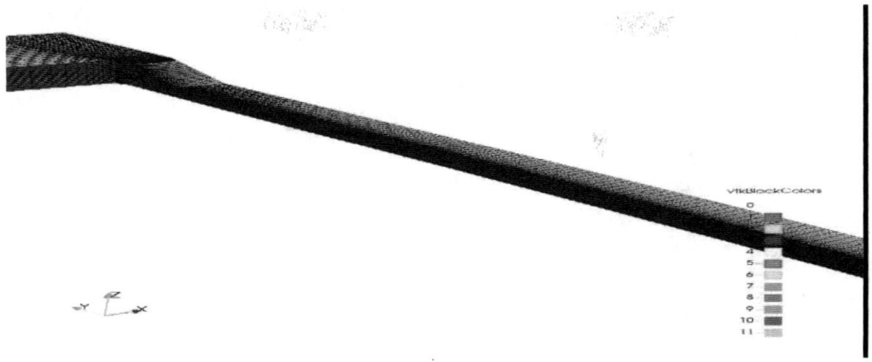

Figure 5. Model Domain

3. Results and Discussion

The comparison between the physical model and numerical model result is presented in Figure 6. The computational mesh is taken as a rectangular grid. The initial condition at the reservoir is set as 40 cm. Measurement point are taken at 2E5 (+1.45 m), 3E5 (+2.45 m), 4E5 (+3.45 m), 5E5 (+4.45 m), 6E5 (+5.45 m), 7E5 (+6.45 m) and 8E5 (+7.45 m).

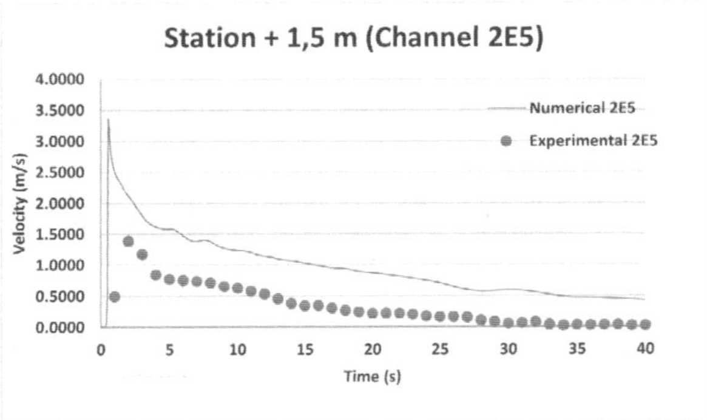

(a)

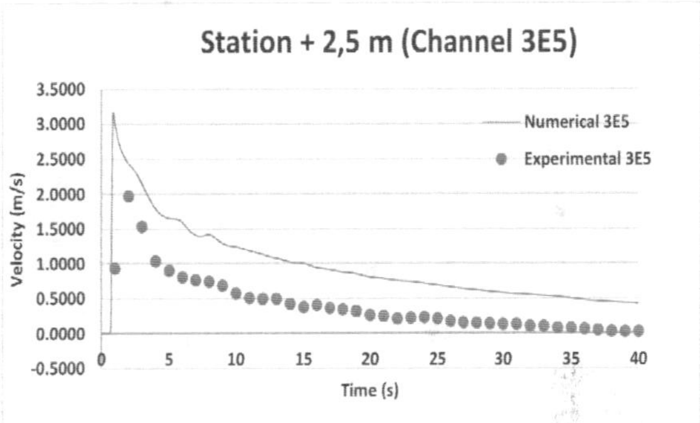

(b)

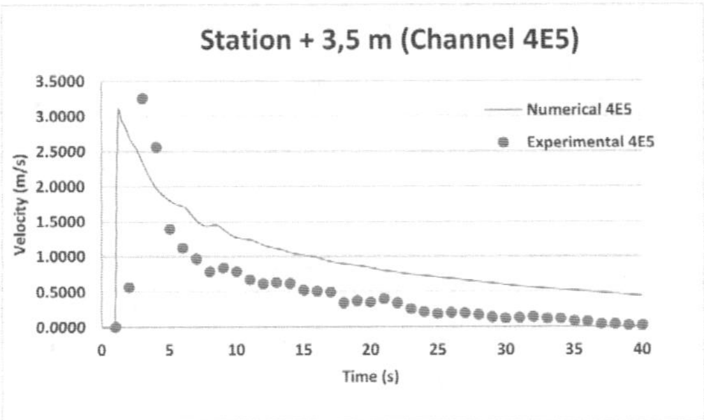

(c)

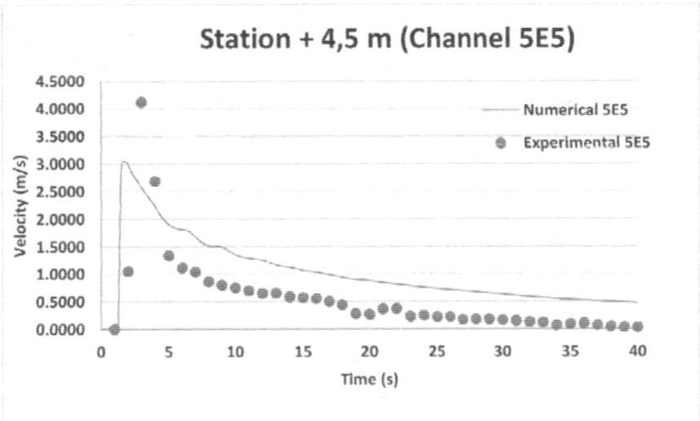

(d)

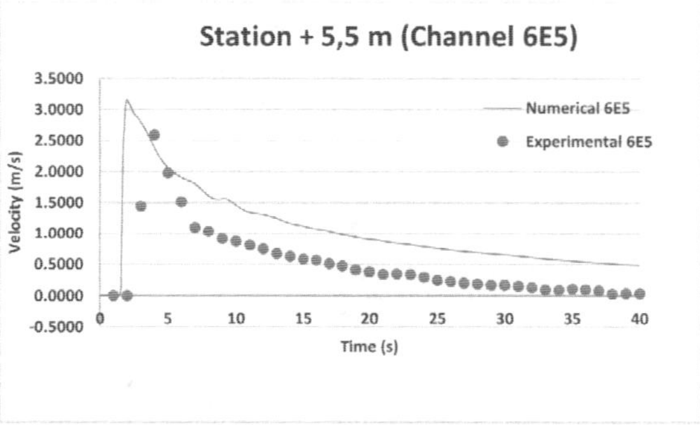

(e)

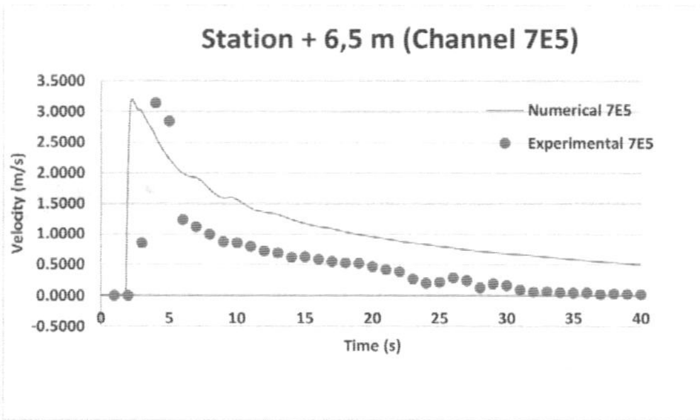

(f)

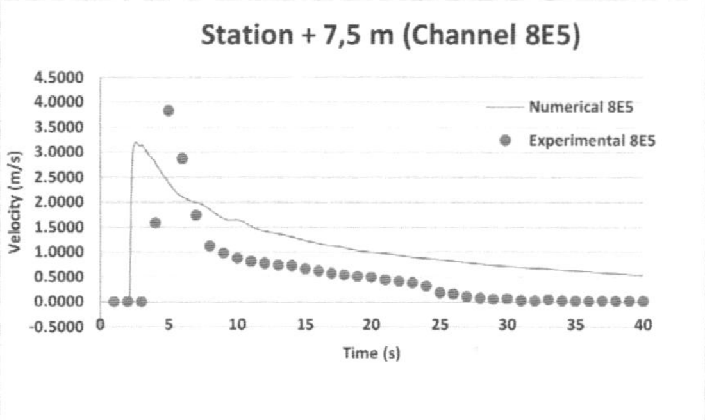

(g)

Figure 6. Time Evolution of Speed for Configuration without Building. (a) Station +1,5 m (Channel 2E5) (b) Station +2,5 m (Channel 3E5) (c) Station +3,5 m (Channel 4E5) (d) Station +4,5 m (Channel 5E5) (e) Station +5,5 m (Channel 6E5) (f) Station +6,5 m (Channel 7E5) (g) Station +7,5 m (Channel 8E5)

Figure 6-a and Figure 6-b show that the numerical model gives a bigger value than the physical model for all time. The comparison results show that there is a velocity difference, an average of 0.5 m/s during the observation time but obtained a similar pattern. Figures 6-c, 6-d, 6-e, and 6-f show the numerical model gives a slightly different result with the physical model, especially for 1-5 s. After 8 s, the numerical still show velocity difference but still have a similar pattern. Figure 6-g shows that at in seconds to 6 and to 7 physical model results indicate that the velocity is greater than the velocity obtained in numerical model results, but after 8 s the numerical result gives the bigger value than the physical model. The results of this study provide the same results as the research conducted by Kusuma et al. [14], wherein a configuration without building the highest speed at the 5th second occurs at 3.5 meters from the gate.

Overall, the value of the numerical model velocity is greater than the value of the physical model. This is because in the numerical model, the conditions are considered ideal, the channel is considered frictionless. Whereas in the physical model, although the bottom of the channel is frictionless, there may be friction due to the bottom condition of the channel even though it is small.

4. Conclusion

The simulation result described earlier shows that a numerical model based on the open-source computational fluid dynamics (CFD) platform called Open FOAM give a good result in solving the dam break problem. Although not giving exact results exactly as resulted in the physical model, the result of this numerical model is able to give an idea of the result of speed changes due to breaking dam flow. In general, there is reasonable agreement between the experimental data and numerical results. Due to the difficulty of obtaining field data for such a flow, laboratory data may be necessary for researchers to verify their mathematical models.

References

[1] Molu, M. 1995. Dam Break Flood in a natural channel: A case study. MS Thesis, Civil Engineering Department, Middle East Technical University, Ankara, Turkey.

[2] Yanmaz, A. M., and Beşer, M. R. 2005. On the Reliability--Based Safety Analysis of the Porsuk Dam. *Turkish Journal of Engineering and Environmental Sciences*, **29**(5), pp.309-320.

[3] Kumar, S., Jaswal, A., Pandey, A., and Sharma, N. 2017. Literature review of dam break studies and inundation mapping using hydraulic models and GIS. International Research Journal of Engineering and Technology, **4**(5), pp.55-61.

[4] Selahatin, K., Hasan, G., Stefania, E., Hatice. 2017. The Influence of Tailwater Depth on 3D Dam Break Wave propagation in an Enclosed Domain, Advamced in Environmental and Geological Science and engineering.

[5] Elkholy, M., LaRocque, L. A., Chaudhry, M. H., and Imran, J. 2016. Experimental investigations of partial-breach dam-break flows. *Journal of Hydraulic Engineering*, **142**(11), p.04016042.

[6] Schmidgall, T., and Strange, J. 1961. Floods resulting from suddenly breached dams: Conditions of high resistance. Hydraulic model investigation, Army Engineer Waterways Experiment Station, Vicksburg, MS.

[7] Miller, S., and Hanif Chaudhry, M. 1989. Dam-break flows in curved channel. *Journal of Hydraulic Engineering*, **115**(11), pp.1465-1478.
Aziz, M. 2000. Dam breaching model. Proc., Joint Conf. on Water Resource Engineering and Water Resources Planning and Management 2000, ASCE, Minneapolis, **56**.

[8] Soares Frazao, S., Le Grelle, N., and Zech, Y. 2002. Dam-break flow experiments in idealised representation of complex topography and urban area. In 1st Project Workshop.

[9] Soares-Frazão, S., and Zech, Y. 2007. Experimental study of dam-break flow against an isolated obstacle. *Journal of Hydraulic Research*, **45**(sup1), pp.27-36.

[10] Lauber, G., and Hager, W., Experiments to dam break wave: Horizontal channel." J. Hydraul. Res., 36(3), 291–307. 1998a

[11] Lauber, G., and Hager, W. H. 1998b. Experiments to dam break wave: Sloping channel. *J. Hydraul. Res.*, **36**(5), pp.761–773.

[12] Samet, K., Ashrafi, M., Golmohammadi, H., and Vatan. M.N. 2019. Shahar Chay Dam-Break simulation using OpenFOAM, 6th International Conference On Civil, architectural, and environmental sciences, Stockholm, Sweden.

[13] Kusuma, M.S.B., Setiawati,T and Farid, M. 2019: Experimental model of dam break flow around several blockages configurations, *International Journal of GEOMATE*, **16**(58), pp.26 – 32.

[14] Pratiwi, V., Kusuma, M.S.B.K., Kardhana, H., and Farid, M. 2020. Experimental Study of Dam Break Flow Generated by the Flap Gate in Horizontal Channel, *International Journal of GEOMATE*, **19**(75), pp.19-26.

Line Learning Software to Improve Learning Interest

Juanda[1*], I E Wessels[2], L Amelia[3], Irmayanti[4]
[1]Departemen Sastra Inggris, Universitas Komputer Indonesia, Indonesia
[2]Departemen Ilmu Komunikasi, Universitas Komputer Indonesia, Indonesia
[3,4]Departemen Manajemen Informatika, Universitas Komputer Indonesia, Indonesia

Email: *juanda@email.unikom.ac.id

Abstract. The internet's role has become crucial. Even when the current Covid-19 is occurring, internet use cannot be avoided. One of the most effective internet services during this pandemic is in education, wherewith the internet can make interactions between teachers and students remotely. Therefore, the purpose of this research is to design online learning applications in Indonesia. This research used the Waterfall method. This method explains the needs of application software in the process of maintaining this application. This application acts as a platform that can help teachers and students. This application provides the teaching staff the freedom to give the material that all students can access freely. This study's result is an online learning application created as efficiently and effectively as possible for use by teaching staff and their students.

1. Introduction

Education is essential for everyone. Education is learning that is usually done directly. During the Covid-19 period, learning experienced several obstacles: students could not understand or interact directly with a school as usual. Therefore, this is where the internet's role becomes essential because it can make interactions between teachers and students remotely. This study aims to design or create online learning applications in Indonesia to increase interest in learning further and facilitate learning. The method for making this application will use the Waterfall Method.

In general, an application is a software or program created and developed to perform specific tasks on a computer, laptop, or smartphone already available for use by users. According to the expert, that application uses a computer, instructions or statements arranged so that the computer can process input into output [1]. Learning comes from the word "leornung", which means an activity of self-development through practice, based on one's ability under the teacher's guidance. While the name "learning implies an effort to teach a person or group of people through various efforts and various strategies, methods approach against achieving the goals that have been planned [2]. Online Learning uses an interactive Internet-based model, such as Zoom, Google Meet, Google Drive, and others [3]. Learning interest is the person's driving force to carry without learning activities to upgrade skills, experience, and knowledge. The interest grows because the desire to understand and know something directs and encourages students' interest in learning to be more serious in learning [4].

The development of information and communication technology has significantly influenced the teaching and learning process in the present Industrial era. Then teachers to facilitate the learning process have used access to technology. Access to technology can also improve the quality of education [5]. It encourages researchers to conduct a study on the use of technology in education, one of which is the title of research that applies access to technology in the field of education, which is mobile-based learning applications for literacy disabilities [6]. This study aims to make it easier for the Deaf to carry out Learning by using the researcher's application [7]. This research explains that there are some basic rules to create a good interface for users. These basic rules are categorized into eight "golden" rules, namely: 1) Enable frequent users to use shortcuts; 2) Fight for consistency; 3) Design dialogue to yield closure; 4) Prevent errors; 5) Offer informative feedback; 6) Permit easy reversal of action; 7) Reduce short-term memory burden; and 8) Support internal locus of control [8]. The researcher uses two design stages to design the application, namely the Splash Screen Interface Design and the Main Menu Interface Design [9]. The research concluded that this application was the eight golden rules of Shneiderman and Plaisant, standard guidelines that can be used to design or build application interfaces for novice-level users or novices (in this case, illiterate users). This learning application is easy enough to be used by illiterates [10].

Education is essential for everyone. Education is learning that is usually done directly. During the Covid-19 period, learning experienced several obstacles: students could not understand or interact directly with a school as usual. Therefore, this is where the internet's role becomes crucial because it can make interactions between teachers and students remotely. This study aims to design or create online learning applications in Indonesia to learn further and facilitate Learning. The method for making this application will use the Waterfall Method.

2. Method

The research used the Waterfall method. It is composed of several phases where each of the steps is defined by different goals and tasks. If stage 1 has not been completed in the Waterfall Method, then stage 2 cannot be initiated, and so on. All phases are interrelated, and each process should be documented and detail.

The Waterfall method requires that every requirement, system objectives, and specification defined in detail. In addition, the waterfall method does not accommodate changes in the middle of the development process. Therefore, the client and analyst team's acquisition in the early stages is the target and the result. Everything should consistent and conform with the acquisition until the application is completed and surrender to the client.

The following are the stages in the Waterfall Model (see Figure 1), namely:

1. This stage gathers the whole needs then analyzed them to define the conditions that the program must fulfill.
2. Design, the result of this stage, produces an overall system design and specifies the software's flow to a detailed algorithm.
3. Implementation, this stage converts the design into program code. The resulting code is in the form of modules that will be implemented into a complete system.
4. Testing, this stage merges the modules that have been made, and tests are carried out to determine whether the software is in accord with the design and whether the software has any errors.
5. Maintenance, this stage is the installation and repair process of the system as approved.

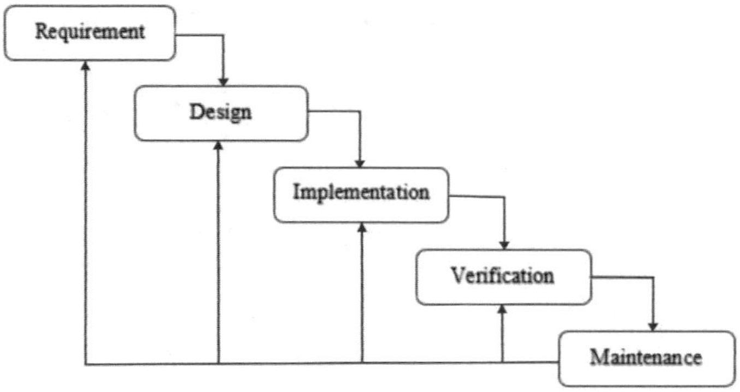

Figure 1. Waterfall Diagram (Data from iq.opergenus.org)

3. Results and Discussion

This section will discuss the analysis of the needs of the pupil's eye. This system will be explained according to the requests requested by teachers and students.

3.1 Analysis

Based on the implementation of the theory put forward by Jirasak that the results of interviews and observations, there are 2 actors in the application of this application, namely teachers and students. It is in Table 1 Actor Identification.

Table 1. Actor Identification

Not	Actor	Description
1	Teacher	People who explain students' eyes to be shared with students in the application.
2	Student	Visitors as students study the subjects in the application.

3.2 Design

This menu structure design explains the relationship between pages and other pages from one menu to another, displaying the application. The method of the farmer's market application is shown in Figure 2 Marketplace application menu structure.

The home menu is the starting page used to view all the menus contained in an application divided into 3, namely the login menu, subjects, practice questions. A page must be filled in by filling in data in the login menu, starting from entering the name, email, and phone number to log in. In the course menu, students can choose what subjects they want to take from various existing issues. Whereas in the question practice menu, students can do practice questions that have been provided with different topics.

3.2.1. Mockup page

Learning applications play an essential role in online learning today. An essential part of the application before it is made is in terms of design that is then given a visual effect so that the image looks its original form to support what the user wants.

Figure 2, the home design or start page, has three menu options: the login menu, the subject menu, and the question practice menu.

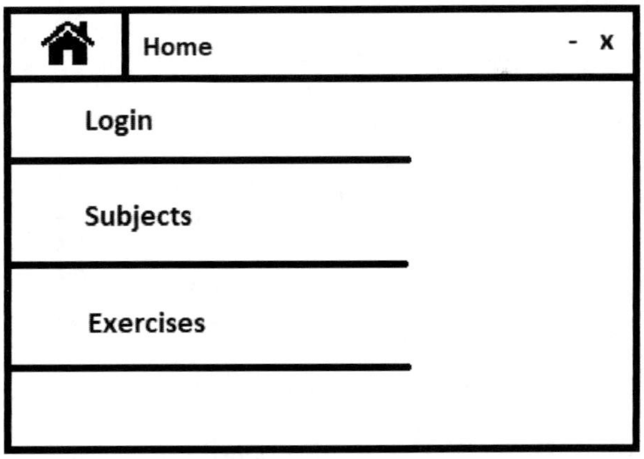

Figure 2. Start Page

Figure 2, Login page interface design for this system is that users must have an account, which at least consists of name, email, password, and number (see Figure 3).

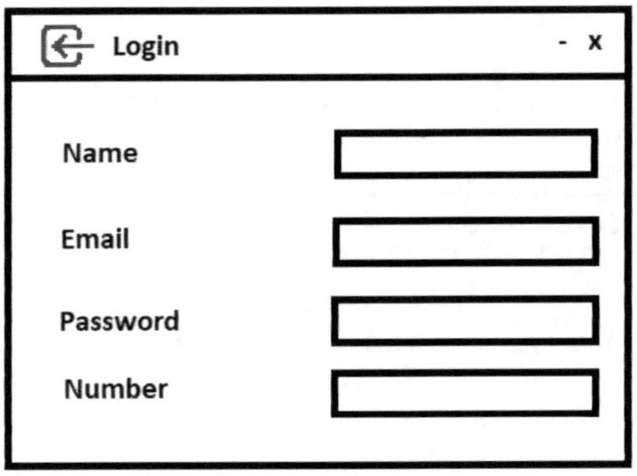

Figure 3. Login Menu

Figure 3, subject page interface design. This page appears after the user logs in, subjects that the user can follow. In addition to the topics that are being followed, a section below displays the user's issues (see Figure 4).

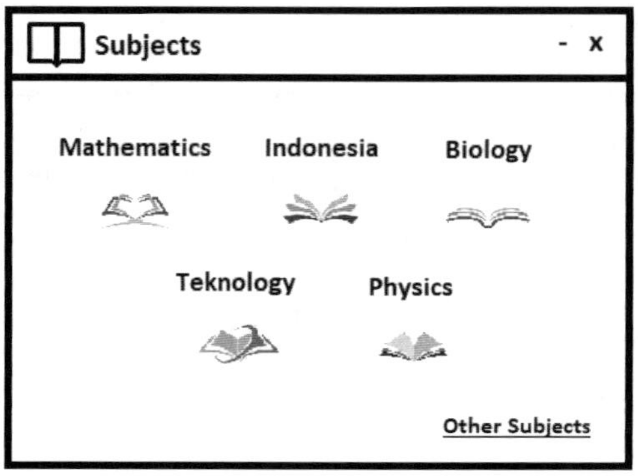

Figure 4. Subject Menu

Figure 4, practice questions page. Design practice questions for this package, the list of practice questions can be seen from the banner, start and end dates if it is still opened (see Figure 5).

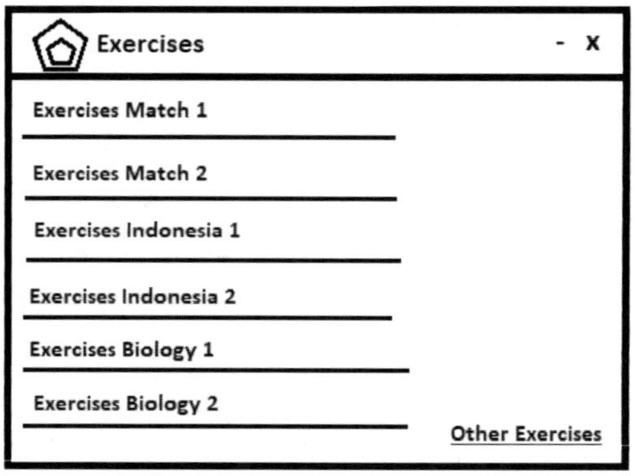

Figure 5. Exercise Menu

4. Conclusion

The education sector is currently having difficulties in learning during the Covid-19 pandemic. Almost all levels of education carry out online learning, namely online-based learning. This is done to prevent the spread of Covid-19, which continues to increase. Online learning also needs an incentive to increase interest in learning for students. With the emergence of this application, it is hoped that students will be more active in carrying out Online Learning during the Covid-19 era.

References

[1] Yani F., Irfan, F., & Mia Z. S. 2020. Motivasi Belajar Mahasiswa Pada Pembelajaran Daring Selama Pandemik Covid-19. *Journal Education*. **6**(2) pp. 166-167.
[2] Mather M., & Sarkans A., 2018 Student Perceptions of Online and Face-to-Face Learning. *International Journal of Curriculum and Instruction*. **10**(2), pp. 61–76.
[3] Muhammad I. P. N., Septiana D. A., 2016. Aplikasi Pembelajaran Berbasis Mobile Untuk Tuna Aksara. *Journal MATICS*. **8**(1), pp. 12-16.
[4] Jirasak S., K., 2014. Developing of Indicators of an ELearning Benchmarking Model for Higher Education Institutions. *Journal Educational Technology*. **13**(2), pp.. 150-152.

[5] Wicaksono A., 2019. Mencari Karakter Pendidik Yang Ideal Bagi Indonesia (Dalam Cerita dan Realita dari Masa ke Masa). *Journal Ilmiah Kependidikan*. **1**, pp. 109-118.
[6] Komalasari, K., Arafat, Y., & Mulyadi, M. 2020. Principal's Management Competencies in Improving the Quality of Education. *Journal Social Work and Science Education*. **1**(2), pp. 181-193.
[7] Wawan, M. Nurul & Waslaluddin, 2014. Analisis Penerapan Sistem E-Learning FMIPA UPI Menggunakan Technology Acceptance Model (TAM). *Jurnal Pengaiaran MIPA*. **19**(1) pp. 129.
[8] Dhull, I., & Sakshi, 2017. Online Learning. International Education & Research Journal (IERJ). *Journal Education*. **3**(8), pp. 32–34.
[9] Aziz Hussin A., 2018. Education 4.0 Made Simple: Ideas For Teaching. International Journal of Education and Literacy Studies. *Journal Education*. **6**(3), pp. 92-98.
[10] Samir Abou El-Seoud, M., Taj-Eddin, I. A. T. F., Seddiek, N., El-Khouly, M. M., & Nosseir, A., 2014. E-learning and Students' Motivation: A Research Study on the Effect of Elearning on Higher Education. International Journal of Emerging Technologies in Learning. *Journal Education and Technology*. **9**(4), pp. 20–26.

Internet of Things Cyber Security in Digital Era

M Solihat*, R V Wulansari

Departemen Ilmu Komunikasi, Universitas Komputer Indonesia, Indonesia

Email: *manap.solihat@email.unikom.ac.id

Abstract. This study aims to determine cybersecurity security in terms of data transfer in the digital era. This study used a descriptive qualitative approach, namely a literature study. The results of this study are to show that using the internet remains safe and avoids dangerous things. With the positive influence of digital technology on cybersecurity, it can be concluded that cybersecurity is to maintain security from technological crimes. This research is expected to serve as a warning to the public to be careful in using the internet.

1. Introduction

Cybersecurity is important because it can protect corporate and personal data. All companies with digital-based data are highly recommended to pay attention to and use cybersecurity when storing, accessing, and retrieving important information. Studying technology security demonstrates the need for new conceptual and analytical resources and cybersecurity research aimed at the technical and social intersection [1]. Technology security nowadays is not just a technical issue. Lithuanian Cybersecurity Law defines cybersecurity as a tool for legal action. The dissemination of technical and organizational information needs to be taken to detect, analyze, prevent, and respond to a problem in technology. It is described as an activity or event that may cause unauthorized access to the system communication and information, electronic communication networks. It is also described as control systems for electronic communication network processes or industrial processes that control operations to destroy, delete, destroy, or modify electronic information and restrict access to electronic information. It also modifies electronic information, restricts, or withdraws access to electronic information. Moreover, it is possible to use non-public information in electronic format by an unauthorized person. The main objective is to provide fast recovery from electronic communication networks, information systems, or industrial process control systems in the event of a cyber-problem or cyber-attack [2].

Identity theft allows data theft to occur, but it is a potentially more serious nuisance. Thus, victims due to data theft are being forced to change a new password. The more important data that can be stolen, the more difficult it is to prove its identity. It will also take time to recreate the documentation [3]. Security and protection in cyberspace must be improved regarding critical information infrastructure, which is very important for every country's safety and economic welfare. Security of internet users has become an essential aspect of developing new technologies and public policies. Cybercrime must be eradicated safely. Since using technical measures alone cannot prevent crime, effective investigation and prosecution of cybercrime by a law enforcement agency is essential. Many countries and governments are currently enforcing strict cyber securities laws to avoid the loss of important information. People must be trained in cybersecurity to be free from crimes from cyberspace [4].

Cybersecurity is protection connected to the internet, including hardware, software, and data from cyber-attacks. The Arizona Cyber Talent Organization states that cybersecurity is the protection of IT systems [5]. Maintaining cybersecurity is to be able to test and provide developing implementations and implement strategies and report every incident so that it can be prepared to act and track the status of network security. Security techniques in technology must solve problems that endanger technology to overcome cybersecurity [6]. The definition of security is a process to protect physical objects against damage, loss, or unauthorized access to maintain the confidentiality and integrity of information about things and to make object information available regardless of time [7].

Security technology implementation is an option and attempts to protect information assets from network attacks or threats or information systems. Currently, technologies present many securities as a security protection for attacks or threats to information networks. Nowadays, many security technologies present as a security protection for attacks or threats to information networks or systems, including cryptographic systems, IDS, SSL, firewalls, IPSec, antivirus systems, authentication, and others [8]. There is no longer any way to place cyber, and it cannot be treated separately from real-world security. If we look to the reality that society's progressive information is taking place, the number of devices connected to the internet also increases.

It is realistic to expect the emergence of new scenarios that violate individual, corporate, national, and global security. The most fundamental problem is that cyberculture is developed faster than cybersecurity, which further implies that cyberspace is integrally at a specific risk [9]. In Indonesia, cyber can still increase. In addition, the number of internet users in Indonesia is still not evenly distributed because there are still many who cannot use internet devices properly, so that cyber law in Indonesia cannot be implemented. Therefore, Indonesia establishes strategic steps to determine cybersecurity and policy based on the National Security for Homeland Security consistency, the US guideline in overcoming various cyber threats to existing infrastructure, and recovering damage caused by cyber-attacks.

Threats and security attacks on social media have been discussed in the previous research. Moreover, based on that research, we must continually protect and secure our data as social media users. In this research, we aim at providing education to be more careful in using technology in the digital era.

2. Method

This research used a descriptive qualitative method, namely the literature study approach. Later, we will see some of the previous research will be used as a reference regarding cybersecurity in the Digital Age, where crimes on the internet are increasingly facilitated, and security in the world of technology is increasingly being threatened. Therefore, cybersecurity must become the center of technological attention in the current era then look for valid and accurate data regarding cybersecurity. With proper and precise data, hopefully, this research is developing according to the need and can be used as learning material about cybersecurity

3. Results and Discussion

3.1 Results

Crime in IoT can occur through anything with various devices. Moreover, every crime can occur. Table 1 shows the crimes on IoT.

Table 1. Crimes in the IoT

No	Source	Devices	Cybercrime in the IoT
1.	Network	Wireless network router, access points, sensor network	Unauthorized access and data modification
2.	Network parameter devices	Firewalls, servers	Unauthorized access
3.	Web	Web servers, web clients, and social networks	Copyright or intellectual property infringement, cyber defamation
4.	Cloud	Cloud systems	Data theft
5.	End nodes	Game consoles, mobile devices, smart TV's, readers	Distribution of malware, data theft, spamming

From the crimes in the IoT based on Table 1, there are several ways to carry out the required security action on the system to minimize crimes in the IoT. Table 2 shows a list of security requirements that are needed for the system in general.

Table 2. General security requirements for the IoT

No	Requirement	Description
1.	Resilience to attacks	To avoid a single point of failure, the system must adapt itself to the failure node
2.	Data authentication	Data and objects that have been retrieved must be authenticated
3.	Access control	From the data provided, the information provided should be able to implement access controls
4.	Client privacy	The system can maintain the privacy of the information provider or client data to provide security and convenience for customers
5.	User identification	Before using users to use the system, they must perform a validation process
6.	Secure storage	Data that has been stored on the system must be kept confidential and ensure the integrity of sensitive information
7.	Identity management	Link user rights and user restrictions to the system to identify individuals/things in the design and control their access to the system
8.	Secure data communication	Ensure the confidentiality and integrity of data communicated by authenticating peer-to-peer communications and preventing rejection of communication transactions, and protecting the identity of communicating entities
9.	Availability	Denial of access or use of user authorization from unauthorized persons or systems
10.	Secure network access	Only authorized devices can provide a network connection or service access
11.	Secure content	Protect the rights of digital content used in the system with Digital Rights Management (DRM)
12.	Secure execution environment	To protect against malicious applications, it is necessary to implement a secure, managed-code, runtime environment designed
13.	Tamper resistance	When the device falls into the hands of malicious parties, then maintain security requirements and check logically

3.2 Discussion

If there has been a crime in the IoT, a forensic investigation must be carried out. The forensic analysis uses science and technology to investigate and establish a criminal or civil court [11]. Forensic investigations aim to carry out experiments to find facts and create the truth of an incident. The greatest challenge for computer forensics is developing methods and providing validation of the techniques, which can be used to produce tangible and reliable evidence to protect against harm [12]. IoT poses challenges to forensic researchers. These challenges are when the crime of spreading data has widened, the lines between networks are blurred, and the privacy of users with personal networks is increasingly fading. Then, it becomes non-private, and private networks become public networks. There are differences when doing traditional and IoT digital forensic investigations. Table 3 shows a traditional digital forensic investigation.

Table 3. Traditional digital forensics investigations

No	Criterion	Traditional digital forensics investigations
1.	Speed of response to incident	Conduct investigation when new events occur
2.	Variety of evidence sources and types	There is sufficiently measured and extensive evidence
3.	Frameworks adaptable	There is a possibility
4.	User input	The perpetrator or victim is usually the user - does no play a role in the investigation
5.	Evidence sources	PC, cloud, web clients, mobile devices, social network
6.	Jurisdiction	Government, company, society, individual, social network
7.	Number of devices	There are billions of devices
8.	Types of evidence	Standard files format and electronic documents
9.	Types of network	Wi-Fi, wireless internet, wired, Bluetooth networks, mobile communications
10.	Quality and type of data and evidence	Up to terabytes of data
11.	Protocols	Wireless, IPv4, IPv6, Ethernet, and Bluetooth
12.	What to seize	As required
13.	Ownership	Governments, companies, individuals, groups

14.	Network boundaries	Relatively clear boundaries s

Then, the IoT digital forensic investigation is shown in Table 4.

Table 4. IoT digital forensics investigations

No	Criterion	IoT digital forensics investigations
1.	Speed of response to an incident	Quite a long response
2.	Variety of evidence sources and types	Sources and evidence that have a more comprehensive range
3.	Frameworks adaptable	Needs frameworks adaptable
4.	User input	Users must remain active using personal IoT, forensic surveillance by the use of adaptable
5.	Evidence sources	Readers, embedded systems sensor nodes, sensor networks, medical implants in humans and animals, other IoTware
6.	Jurisdiction	Government, company, society, individual, social networks
7.	Number of devices	From 50 billion to trillions of devices
8.	Types of evidence	Allows all formats
9.	Types of networks	Sensor network, sensor to reader and vice versa, RFID
10.	Quantity and type of data as well as evidence	Up to exabyte of data
11.	Protocols	RFID
12.	What to seize	Identify the best thing for the source the following evidence
13.	Ownership	Governments, companies, individuals, groups
14	Network boundaries	Boundary lines that have no clarity

With the challenges in carrying out digital forensic investigations on IoT, data security on IoT is much needed; it meets the system's general security requirements that users will use. It meets the general conditions of security on the design and can also guarantee user data confidentiality.

4. Conclusion

Technological development can make it easier for people to get information quickly. The threats and attacks on technology nowadays are being the most worrying things. In this digital era, it is necessary for us to always be careful in protecting all data because many crimes are committed in any way. Along with the development of technology, the threats will also get heavier. Therefore, there is a need for security to safeguard data.

Acknowledgment

We want to thank Prof. Dr. Ir. H. Eddy Soeryanto Soegoto, MT as a lecturer in Entrepreneurship in the Field of Information Systems and Communication at Universitas Komputer Indonesia, who has provided knowledge and insight for this research.

References

[1] Liebetrau, T., & Christensen, KK 2020. The ontological politics of cybersecurity: Emerging agencies, actors, sites, and spaces. *European Journal of International Security*, pp. 1-19.
[2] Limba, T., Plėta, T., Agafonov, K., & Damkus, M. 2019. Cybersecurity management model for critical infrastructure.
[3] Arlitsch, K., & Edelman, A. 2014. Staying safe: Cyber security for people and organizations.*Journal of Library Administration*, **54**(1), 46-56.
[4] Iqbal, H., Al-Utaibi, G., & Bohra, O. P. The Reality of Technologies for Cybersecurity Challenges.
[5] Morantz, B. Cyber Security Is an Application of Decision Sciences.
[6] Callen, J., & James, J., E. 2020. Cybersecurity Engineering: The Growing Need. *Issues in Information Systems*, **21**(4).
[7] Abomhara, M., & Køien, GM 2015. Cyber security and the internet of things: vulnerabilities, threats, intruders and attacks. *Journal of Cyber Security and Mobility*, pp. 65-88.
[8] Bustami, A., & Bahri, S. 2020. Threats, Attacks and Protection Measures on Network or Information System Security: Systematic Review. *UNISTEK*, **7**(2), pp. 59-70.
[9] Mihailović, A., & Rašović, N. Cybersecurity in the New Reality: Systematic Review in the Context of COVID-19.
[10] Simbolon, IS 2017. Cyber Initiatives in the Context of Cybersecurity in the Philippines.*Asymmetric Warfare*, **3**(2).
[11] Philipp, A., Cowen, D., & Davis, C. 2009. *Hacking Exposed Computer Forensics: Computer Forensics Secrets & Solutions*. McGraw Hill Professional.
[12] Noblett, M. G., Pollitt, M. M., & Presley, L. A. 2000. Recovering and examining computer forensic evidence. *Forensic Science Communications*, **2**(4).

Designing Digital Certificate Issuance Information System

D Albar[1], B F F Perdana[2]
[1]Departemen Desain Komunikasi Visual, Universitas Komputer Indonesia, Indonesia
[2]Fakultas Teknik dan Ilmu Komputer, Universitas Komputer Indonesia, Indonesia

Email: deni.albar@email.unikom.ac.id

Abstract - Nowadays, some training and webinars are conducted frequently. Especially during the Covid-19 pandemic, online training and webinars became rational choices in the education sector. Evidence of participation from training and webinars is usually a digital certificate. Manual digital certificate creation will usually take a lot of time. It will be easier if the process is made systematically. This research aims to design a website-based application that can publish digital certificates that event participants can download. The method used in designing and making this website is descriptive. It is expected to present a complete picture of the subjective or objective of the related phenomena. The result of this research is a website-based application design to make it easier to issue digital certificates. Automation of the system can help simplify work and broad accessibility due to the website platform's use. From the results obtained, it can be concluded that by using this application, making certificates utilizing the system is more efficient than the manual method and can reduce the amount of paper usage. With this application, it is hoped that it will make it easier for the event committee to process the issuance of digital certificates and make it easier for participants to get their certifications.

1. Introduction

Nowadays, when a pandemic occurred in Indonesia and other countries in 2020, many activities have shifted from offline to online. Seminar activities usually held in the community, in the educational and non-educational world, which are typically held onsite, have been converted into online activities. Seminar activities, which are usually held by gathering many people in one place, were blocked by the pandemic, replaced by webinars. Webinars are one of many training methods. According to Gegenfuther and Ebner, webinars are defined as web-based seminars, where students and teachers are connected virtually together and interact in real-time [1]. Currently, course service institutions that provide specific competency courses online hold many webinars. In each of these online competency course activities, usually, this course service offers certificates to the participants, presenters, and committee [2]. A certificate is a documentary proof of a person's participation in the event [3]. An electronic certificate contains an Electronic Signature that shows the status of a legal subject issued by the Operator [4]. Digital certificates are used because, currently, seminars are being held online. After all, electronic certificates can be used [5]. Manual digital certificate creation will usually take a lot of time; it will be easier to make the process in a systemized manner. The use of a QR-code, which stands for “Quick Response,” is a matrix barcode containing text data or a URL that can be used to verify the digital certificate's authenticity [6].

Several studies on digital certificate creation have been done before. For example, research by Nureni et al. concluded that making digital certificates requires the use of QR-codes on electronic certificates to create the certificate verification process simpler [2]. It is supported by Erick's research, which concludes that QR-Code can be used as a certificate verification system [7]. It is also proven by the emergence of an average SUS score of 83.33 from questionnaires taken from 40 respondents [7]. For further research by Rafi, this study describes more complicated and safer security with blockchain technology [8]. Based on previous research, research regarding designing the printing system of the digital certificate. Most of the previous research focused on certificate security using QR-code as well as blockchain. Meanwhile, this research is focused on designing an information system that supports the online digital certificate printing process.

2. Method

This research focused on designing a digital certificate printing system and using QR-code technology to support the verification of the authenticity of certificates using the Laravel framework at PT. Someah Kreatif Nusantara. The method used in this research was the descriptive method. It is used by presenting subjectively or objectively a situation description related to business architecture planning and planning and development of information systems. In the end, a definition of the linkages and benefits of Business Architecture Planning (EAP) is found with information system development. The workflow of the Digital Certificate Issuance information system can be illustrated as shown in Figure 1.

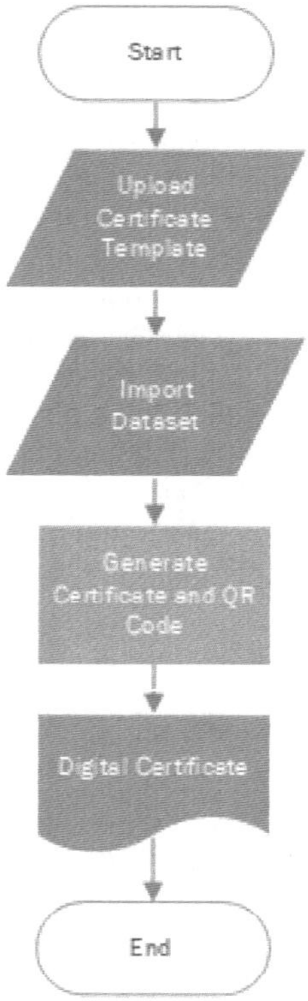

Figure 1. Digital Certificate Issuance Workflow.

This information system has three main processes: uploading the certificate template, importing the dataset, and generating the certificate template and the dataset into a digital certificate in a .pdf file. This information system is created with the Laravel framework, making website development easier, elegant, simple, and fun because of Laravel's excellent features such as the Template Engine, Routing, and Modularity [9].

3. Results and Discussion

3.1. Designing Certificate Template Design

This digital certificate issuance system's flow starts with creating a template using the Microsoft Word application with a background made with a graphic editing application. In this section, several variables must be entered so that the certificate results match each person's data (datasets). This template depends on the certificate design that will be printed (see Figure 2).

Figure 2. Template Digital Certificate

In Figure 2, if the template is created, variables are inserted, which will be printed based on each dataset. Therefore, each dataset will display different data from one another with the same template.

3.2. Interface

The user interface is one of the essential parts of the application. The user interface must satisfy user needs. Appropriate interface design must meet users' needs, capabilities, and limitations in using the information system [10]. On the front view, the webinar participants are presented with the display as shown in Figure 3.

IOP Conf. Series: Materials Science and Engineering 1158 (2021) 012018 doi:10.1088/1757-899X/1158/1/012018

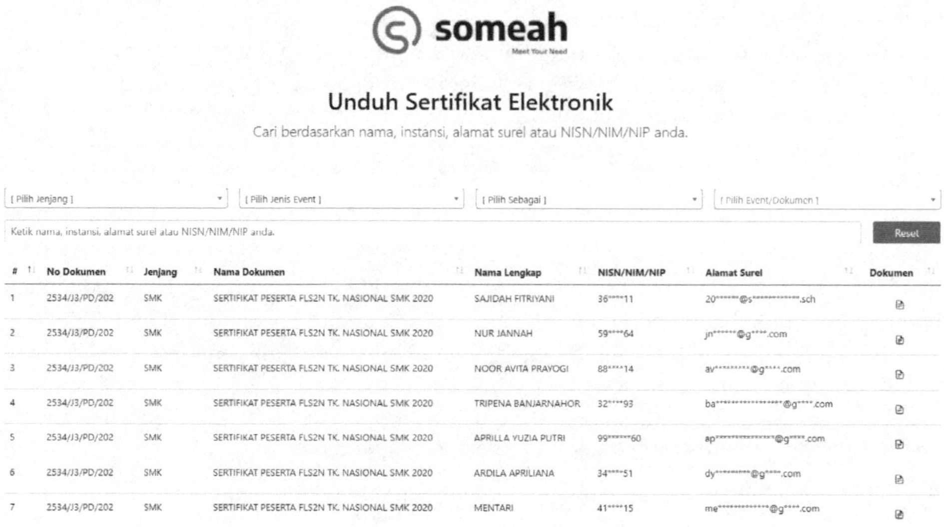

Figure 3. Participant User Interface

Meanwhile, the administrator interface is more complex because it requires an interface that can cover the data requirements as are necessary for issuing digital certificates. It is shown in Figure 4.

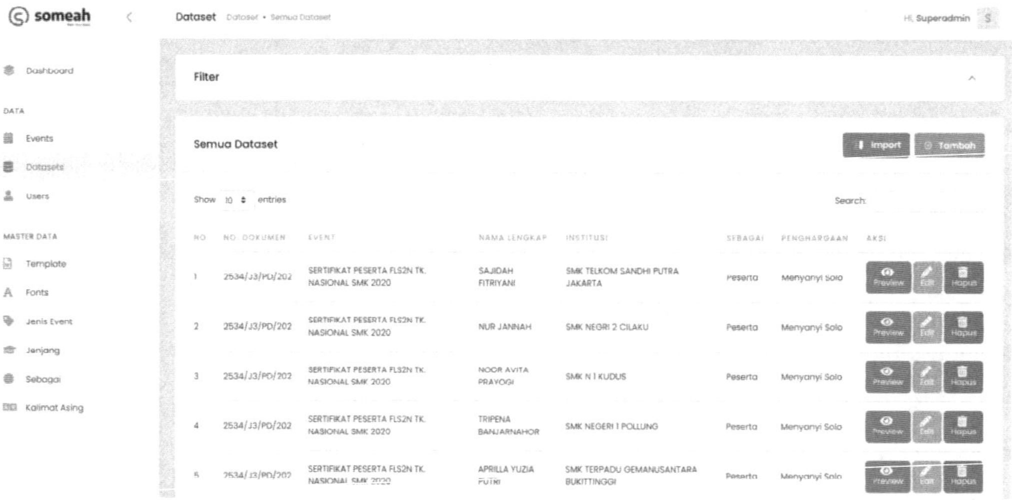

Figure 4. Admin User Interface

3.3. Digital Certificate Printout

Each registered participant can download the results of the digital certificate printing process (see Figure 5).

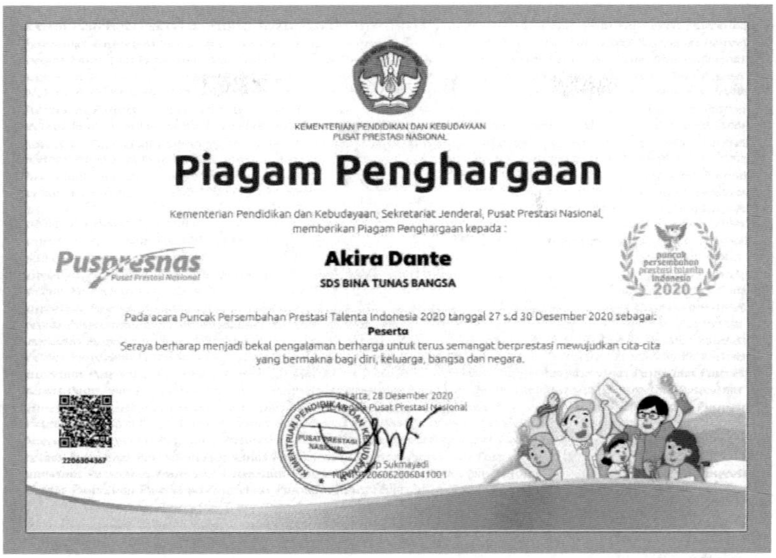

Figure 5. Digital Certificate Printout

In Figure 5, each registered participant can download the digital certificate printing process results, and the webinar organizer inputs the data.

3.4. *Performance Benchmarks*

In measuring the Digital Certificate Issuance information system's performance using Google Lighthouse, the following performance is obtained (see Figure 6).

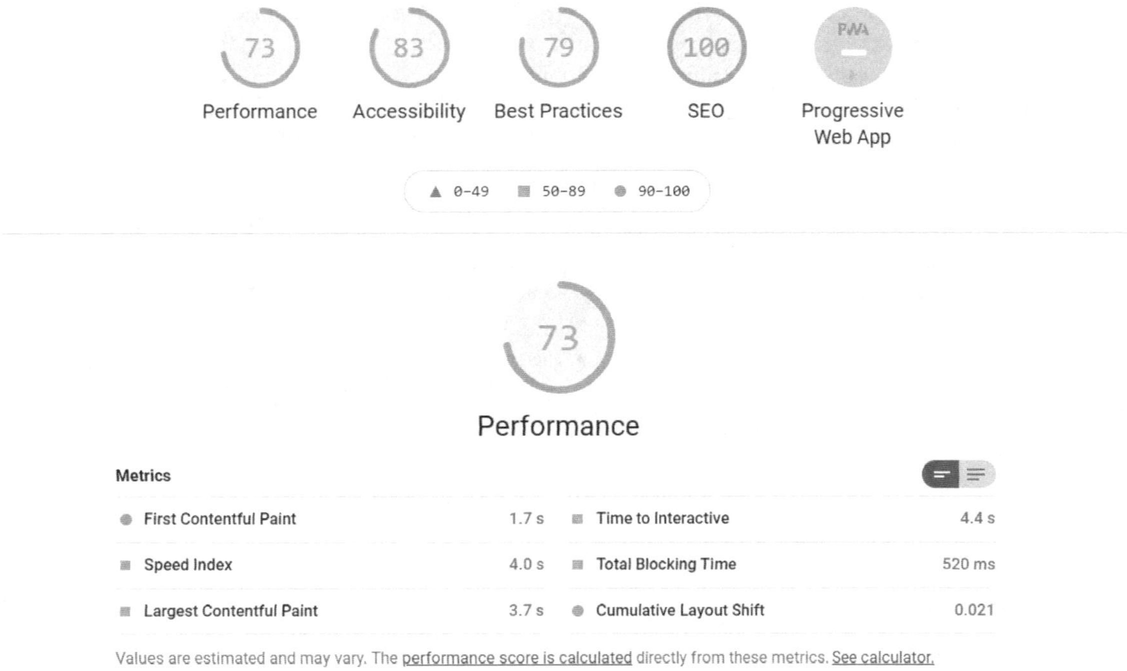

Figure 6. Performance Benchmark with Google Lighthouse

Using the Laravel framework and several plugins such as QR-Code generators and converters from Docx to Pdf, the results are promising, with an average value above 70. This result can describe the information system's performance regarding the excellent or bad technical implementation of the information system. Yu stated that the experimental and simulation proved that web design based on a Laravel framework has scalability and robust scalability to improve the developing efficiency [11].

4. Conclusion
In conclusion, it is found that using certificate templates can be helpful and easy because they can be reused on other people's certificates. The use of a QR-code can be a means of validating a certificate that has been issued so that the certificate can be checked for the authenticity and conformity of the information stated on the digital certificate. For further development, it is recommended to make comparisons on blockchain as a form of security and data validity from digital certificates. Therefore, we can get the truth or not the use of blockchain technology in digital certificate certificates

5. References
[1] Gegenfurtner, A., Zitt, A., & Ebner, C. 2020. Evaluating webinar-based training: a mixed methods study of trainee reactions toward digital web conferencing. *International Journal of Training and Development*, **24**(1), pp. 5-21.

[2] Nuraeni, F., Agustin, Y. H., Kurniadi, D., & Ariyanti, I. D. 2020. Implementasi Skema QR-Code dan Digital Signature menggunakan Kombinasi Algoritma RSA dan AES untuk Pengamanan Sertifikat Elektronik. In *Seminar Nasional Teknologi Informasi Komunikasi dan Industri* p. 43.

[3] Gokulakrishnan, S., & Sarma, C. S. 2020. An Approach to E-Certificate Designing with Auto-Emailing. *International Journal of Scientific Research & Engineering Trends*, **6**(3).

[4] Somsuk, K., & Thakong, M. 2020. Authentication system for e-certificate by using RSA's digital signature. *Telkomnika*, **18**(6), pp. 2948-2955.

[5] Gopal, N., & Prakash, V. V. 2018. Survey on Blockchain Based Digital Certificate System. *International Research Journal of Engineering and Technology (IRJET)*, **5**(11).

[6] Ho, C. T. B., & Yang, J. M. D. 2017. Factors affecting users' mobile technology usage intentions: an example of QR code scanning for mobile commerce. *International Journal of Mobile Communications*, **15**(2), pp. 185-209.

[7] Febriyanto, E., Rahardja, U., Faturahman, A., & Lutfiani, N. (2019). Sistem Verifikasi Sertifikat Menggunakan Qrcode pada Central Event Information. *Techno. Com*, **18**(1), pp. 50-63.

[8] Alrawais, A., Alhothaily, A., Cheng, X., Hu, C., & Yu, J. (2018). SecureGuard: A certificate validation system in public key infrastructure. *IEEE Transactions on Vehicular Technology*, **67**(6), pp. 5399-5408.

[9] Soegoto, E. S. 2018. Implementing Laravel framework website as brand image in higher-education institution. In *IOP Conference Series: Materials Science and Engineering*. **407**(1), p. 012066.

[10] Ruiz, J., Asensio, E. S., & Snoeck, M. 2020. Learning UI Functional Design Principles Through Simulation With Feedback. *IEEE Transactions on Learning Technologies*, **13**(4), pp. 833-846.

[11] Mahmood, M. T., & Ashour, O. I. 2019. Design and Implementation of Web Based For Intermediate Online Shop with Laravel Framework. *Int. Journal of Comp. Science & Mobile Computing*, **8**(3), pp. 124-133.

IOP Conf. Series: Materials Science and Engineering 1158 (2021) 012019 doi:10.1088/1757-899X/1158/1/012019

The Study of Main Component of the High Temperature Gas-cooled Reactor Balance of Plant using Local Component

Muksin Aji Setiawan[1*], Marliyadi Pancoko[2], Kiswanta[1], Syaiful Bakhri[1]

[1]Center for Nuclear Reactor Safety and Technology, National Nuclear Energy Agency of Indonesia (BATAN), Kawasan Puspiptek Serpong, Tangerang Selatan, 15314, Indonesia
[2]Center for Nuclear Facility Engineering – National Nuclear Energy Agency, Kawasan Puspiptek Serpong, Tangerang Selatan, 15314, Indonesia

*Corresponding author: muksin-aji@batan.go.id

Abstract. National Nuclear Energy Agency (BATAN) as a research and development institution related to nuclear energy has designing Nuclear Power Plant (NPP) as one of the solutions to meet energy needs in Indonesia. This alternative electric power plant can be used as back up of fossil base power plant and maintaining the basic load of electric consumption. This research purpose is to simulate the balance of plant of the designed NPP and comparing the main component need to be produce by local industry. The simulation was created using Chemcad software to describe the process and component analysis of the NPP. The simulations done also can be increase the economic efficiency. Simulation results show the component data which can be used as a reference for the selection of the main components and pipes would be used in the construction of the reactor. The simulation results data of this simulation also can be uses to decide the local component that can apply in future NPP.

1. Introduction

Most energy in Indonesia is currently filled with utilizing fossil fuel. Energy needs in Indonesia continue to increase annually, to solve this problem there must be considering using several new and renewable energy sources. One of best candidate energy source is Nuclear energy. In a Global scale of Indonesia for 2040 Special Volume reviews predicted increase to almost 50 x 1018 J needed for Nuclear energy [1]. The construction of NPP in Indonesia has been planned for a long time by BATAN. The Reactor should be one of the best solutions because the other energy source such as the chemical and liquid fuel production quite complex considering the supply chain [2]. The reactor design carried out by BATAN is currently a fourth generations reactor which is a HTGR. This type of reactor was chosen because of the high safety factor. The reactor designed by BATAN is an Industrial and Steam Power Plant (PeLUIt), this reactor is designed with a power of 150 MW. This reactor was designed to generate electricity and steam to be used for industrial purposes. HTGR research progress in global was started by German many decades ago and recently the research lead by China.

HTGR design as commercial NPP generally choose because of its safety factor. The PeLUIt design offer many utilities when it operates. Before the construction, the design should be validated and verification that the safety standard has been completed. This research would be basic for understanding the process of BATAN design on nuclear reactor. The simulation result also can be used to other simulations in a different arrangement of the design. The model- based simulation generally decided by the accuracy of model represent the real condition in a same tame using simple approaches [3]. The simulation also can be used to enhance the skill of designer to know the error and provide information to the operator in control room after NPP constructs [4]. This research is important because the results of the simulations carried out serve as a reference for the availability of viewing components from current local industry conditions. In the preliminary assumptions, the design would need specific component that must be fabricated because it is not the general market consumption. In further scope the simulation has advantage because it can be more economical efficient than creating real experiment set to know the phenomenon.

The research performing a process simulation that occurred in a steady state. The simulation results are used as data to determine the needs of the components and pipes to be used. The selection of components out based on the calculation of ideal conditions in the circumstances of the operation. For the development simulation can be enhance using web-bases simulation. This kind of simulation has a good point in control from the distance. The simulation can be access and be modify by research team member [5]. Simulator of NPP can be classified to several kind. There is offline and online simulator. Type of simulators and their purpose can be divided to simulator games, Desk-top simulators and control room simulators [6]. There are at least three stages use of simulator studies: (1) systematic studies of natural condition and its complexity (2) full scale simulation which can be adjust by the operator base on condition change (3) the flexibility and low cost to test alternative design [7]. Calculations made are taken into consideration in selecting the component specifications to be used. Equilibrium conditions in the designed reactor system are calculated using the help of Chemcad software. Chemcad is software commonly used in determining the calculation of system equilibrium in equilibrium. In the future most of process design and simulation will be using software that integrated together to solve global optimization problems [8].

2. Methods

The PeLUIt NPP design basically was the modification of the Experimental Power Reactor which is a Non-Commercial Power Reactor (RDE RDNK) that BATAN is working on inspired by HTR10 and HTR PM models. The most obvious PeLUIt design difference is the amount of output power of 10 MW and 150 MW. Unlike the water-cooling type reactor, the High Temperature Reactor (HTR) model has more complex flow and heat transfer with higher turbulence levels. This has led to the need to also calculate the thermohydraulic parts as a consideration [9]. The limits values operation and safety were shown in Table 1.

Table 1. The considered limit values of operation and safety [10]

Parameter	Limit	
	Operation	Safety
Max neutron flux at low power (>15 MWt)	150%	15 MW
Max neutron flux at middle power (15-75 MWt)	110%	120%
Minimum reactor period	45 s	20 s
Max. He outlet temperature	780°C	790°C
Max He inlet temperature	270°C	290°C
Max He flowrate	110%	120%
Max. thermal power increase	2.3%/min	3.5%/min
Max. primary pressure increase	0.01 MPa/min	0.03 MPa/min
Max. primary pressure decrease	0.01 MPa/min	0.03 MPa/min
Max. ratio of He and steam flowrate	1.20	1.30
Max. ratio of steam and He flowrate	1.17	1.33
Max He humidity	50 ppmv	800 ppmv
Max. steam pressure decrease	0.6 MPa/min	1.0 MPa/min
Max. deviation for He flowrate	20%	20%

Simulation has two types, steady and dynamic. The simulation in this study is basically a simulation at a steady state [11]. Simulator design can basically be used for transient conditions and analysis of emergency conditions, verification of design changes, power tests produced, and verification of system design and integration [12]. Mass balance and energy produced can also be calculated. The main purpose in making simulations for the design that is being done is to guarantee in terms of safety. To fulfill this requirement, it is necessary to do a lot of verification and validation of the design that is done. Granting licenses to continue to improve safety and reduce risk in accordance with the new technology used by [13]. The simulation in this research split into two steps, the primary and the secondary system, and how to connect this system. In primary loop, cooling system using gas as coolant and not mix with the water from the secondary loop [14]. Continuous improvement needs to be done with regard to all circumstances so that it can provide an overview of the most effective reactor design in energy and mass balance. In the cooling system section, further research is also needed because the amount of power produced by the power plant is strongly influenced by the success in the cooling process [15]. Combining both primary and secondary loop to create full simulation of the NPP design would describe the process of the balance of plant. The simulation result can be used to decide which component already exist as local component. The flow of the research simulation can be describe as shown in Figure 1.

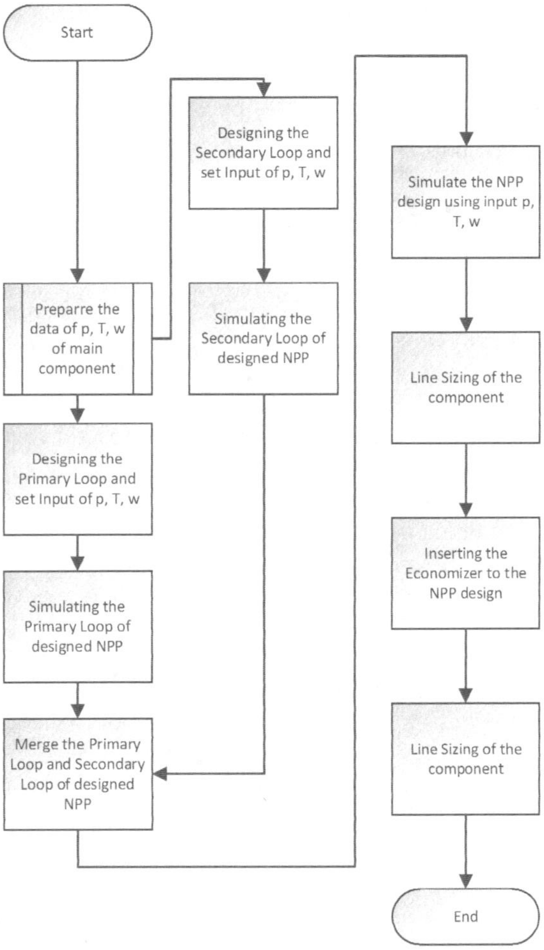

Figure 1. Flowchart of the Research

3. Results and Discussion

Simulation result of the PeLUIt 150 MW after combining the primary and secondary system shown in Figure 2. The simulation describes the process between two loops of fluid inside the cooling system and how the heat generation each loop. There are three main variables for each main component:

temperature, pressure, and flow rate of the coolant. The other variable is not shown Chemcad but recorded by simulation result data sheet.

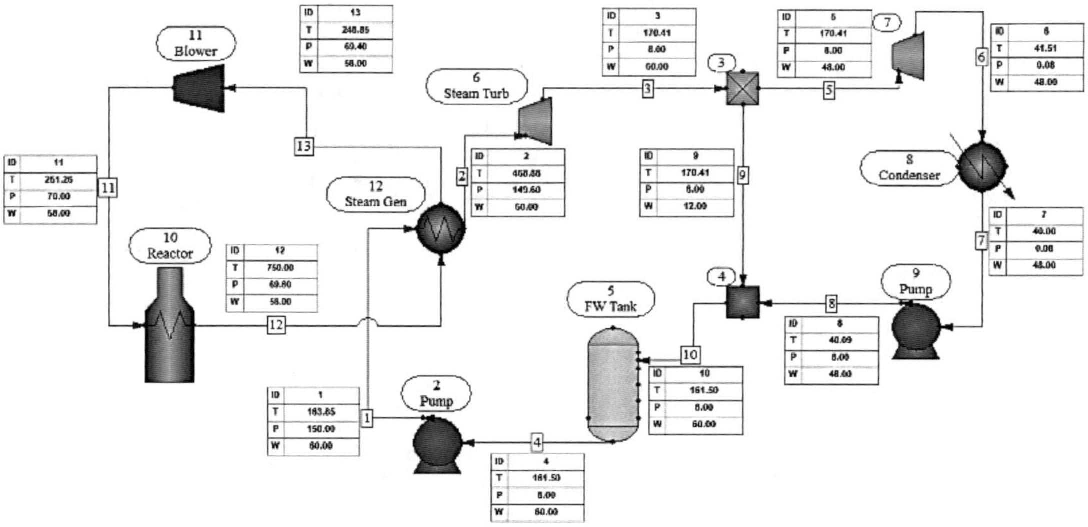

Figure 2. Simulation Results of PeLUIt 150 MW

The simulation adds some virtual component to make a simple running because some limitation of the available component in Chemcad. This system can represent the general design for the 150 MW PeLUIt which consists of a helium gas primary cooling system reactor and a cooling system that uses water. In accordance with the hypothesis in the previous simulation, the simulation results which are a combination of the two previous simulations have finally changed the previous conditions. Some of these changes occur in the magnitude of the magnitudes and constants associated with the amount of energy generated. In the first simulation the heat absorbed amounted to 150.3549 MJ / h finally changed to 150.3427 MJ / h. meanwhile some thermohydraulic parameters are maintained as values in the previous simulation.

The stream properties possessed by this combined simulation show that changes in several degrees of celsius, but there are some parts that do not experience changes in temperature. The phenomenon where enthalpy values change almost in all parts which are components of the system cycle with water flow is interesting. This changing enthalpy value is more or less caused by mixing the flow that occurs in the steam generator with helium flow. It is supported that the molar flow value does not change significantly. The stream properties of PeLUIt 150 MW for each component detailed in Table 2.

Table 2. Stream Compositions of PeLUIt 150 MW

Stream No.	Temp C	Pres bar	Enth MJ/h	Vapor mole fraction	Molar flow kmol/h	Mass flow kg/h	Std liq m3/h	Std vap 0 C m3/h	Water flow rates kg/h	Helium flow rates kg/h
1	163.8462	150	-916.86	0	3.3306	60	0.06	74.6499	60	0
2	468.8775	149.6	-765.78	1	3.3306	60	0.06	74.6499	60	0
3	170.4135	8	-794.71	0.9844	3.3306	60	0.06	74.6499	60	0
4	161.4961	8	-917.96	0	3.3306	60	0.06	74.6499	60	0
5	170.4135	8	-635.77	0.9844	2.6644	48	0.048	59.72	48	0
6	41.51	0.08	-660.08	0.8557	2.6644	48	0.048	59.72	48	0
7	40	0.08	-759.07	0	2.6644	48	0.048	59.72	48	0
8	40.087	8	-759.02	0	2.6644	48	0.048	59.72	48	0
9	170.4135	8	-158.94	0.9844	0.6661	12	0.012	14.93	12	0
10	161.4961	8	-917.96	0	3.3306	60	0.06	74.6499	60	0
11	251.2575	70	69.199	1	14.4906	58	9.816	324.7867	0	58
12	750	69.6	219.54	1	14.4906	58	9.816	324.7867	0	58

4. Conclusion

Simulations for the 150 MW reactor model are made by combining two different fluid streams. Physical phenomena that occur affect the thermohydraulic value in the reactor system. Variations in making simulation schemes are interesting because optimization for the system is the best. Addition of components to the simulation design can affect the balance of plant and mass-energy balance. Simulation that has been done can be used as a reference as the selection of reference components that can be used when the reactor will be built. Simulation results cannot yet be ascertained for determining the availability of components that have been produced locally in Indonesia. The same is true for the selection of pipes to be used. the determination of pipe size can be done directly from the simulation results. Further research is needed regarding the selection of components available in Indonesia whether these components can be used for the purposes of nuclear power plants.

Simulation base on Chemcad for the steady state condition of PeLUIt model process can be use in describe the balance of plant the NPP model which is the scale-up of the RDE-RDNK design by BATAN. However, limitation on the simple simulation model there are some variable not calculated. The manufacture of the component also become other case to solve. Some calculated component doesn't produce for general industrial usage. There are two options to solve this problem, using the new manufacture product or using the existing component but re-arrange the simulation. The better option was created the new simulation that suit to the local industry component that produce in Indonesia. From this simulation, the research show there would be need another study and continuous comparation about the capability of the local industry to support the nuclear plant construction.

Acknowledgement
This work is funded by INSINAS 2020 project entitle "*Pengembangan Simulator Engineering PLTN 150 MWatt Tipe Classroom*"

References

[1] Adams II, T. A. 2015. Future opportunities and challenges in the design of new energy conversion systems. *Computers & Chemical Engineering*, **81**, pp.94-103.
[2] Martín, M., and Adams II, T. A. 2019. Challenges and future directions for process and product synthesis and design. *Computers & Chemical Engineering*, **128**, pp.421-436.
[3] Vajpayee, V., Becerra, V., Bausch, N., Deng, J., Shimjith, S. R., and Arul, A. J. 2020. Robust-optimal integrated control design technique for a pressurized water-type nuclear power plant. Progress in Nuclear Energy, p.103575.
[4] Li, Y., and Mosleh, A. 2018. Dynamic simulation of knowledge based reasoning of nuclear power plant operator in accident conditions : Modeling and simulation foundations *Saf. Sci.*, no. May

2017, pp. 1–15 doi: 10.1016/j.ssci.2018.02.031.

[5] Kim, K. D. 2007. A web-based nuclear simulator using RELAP5 and LabVIEW. *Nuclear Engineering and Design*, **237**(11), pp.1185-1194.

[6] Kerlin, T. W., and Upadhyaya, B. R. 2019. Chapter 15 - Nuclear plant simulators," T. W. Kerlin and B. R. B. T.-D. and C. of N. R. Upadhyaya, Eds. Academic Press, pp. 203–211.

[7] Theureau, J. 1998. Use of Nuclear-Reactor Control Room Simulators in Research and Development, *IFAC Proc.* Vol., **31**(26), pp. 365–370

[8] Alwi, S. R. W., Klemeš, J. J., and Varbanov, P. S. 2016. Cleaner energy planning, management and technologies: Perspectives of supply-demand side and end-of-pipe management. *Journal of Cleaner Production*, **136**, pp.1-13.

[9] Yuanle, M., and Shuyong, Z. 2012. Simulation for the secondary loop of the chinese 200mwe htr-pm base on vpower. *Energy Procedia*, **16**, pp.1831-1838.

[10] Subekti, M. 2019. Brieft Report of The HTGR Simulator Development, PTKRN BATAN

[11] Edwards, J. E. 2013. Process Simulation Dynamic Modelling & Control, *P&I* Des. Ltd

[12] Lu, J., Guo, J., Kozlowski, T., and Li, F. 2018. Coupling methods for HTR-PM primary circuit. *Annals of Nuclear Energy*, **116**, pp.20-30.

[13] Martín, M., and Martín, M. 2017. Cooling limitations in power plants: optimal multiperiod design of natural draft cooling towers. *Energy*, **135**, pp.625-636.

[14] Memmott, M. J., Wilding, P. R., and Petrovic, B. 2017. An optimized power conversion system concept of the integral, inherently-safe light water reactor. *Annals of Nuclear Energy*, **100**, pp.42-52.

[15] Zhou, Y., Zhou, K., Ma, Y., and Sui, Z. 2013. Thermal hydraulic simulation of reactor of HTR-PM based on thermal-fluid network and SIMPLE algorithm. *Progress in Nuclear Energy*, **62**, pp.83-93.

IOP Conf. Series: Materials Science and Engineering 1158 (2021) 012020 doi:10.1088/1757-899X/1158/1/012020

The Analysis Chemical Composition and Microstructure of Bolt Anchor Bushing in Cessna Caravan Type 208B

Indreswari Suroso
Aeronautical Program, Yogyakarta Aerospace Technology College

Email: indreswari.suroso@sttkd.ac.id

Abstract. Bolt Anchor Bushing is a component contained in an aircraft in the form of a cylindrical iron tube in which there is rubber. Bolt Anchor bushing functions as a bearing between moving and immovable components. This is done so that aircraft components can carry out their functions properly and safely. Bolt Anchor Bushing must be sturdy and strong for the shaft elements and the machine to work properly. If the Bolt Anchor bushing does not work properly, it will affect how the component system works. The methods in this research were observation, interviews, quotes with journals, and testing of chemical composition and microstructure. The results of characteristics Bolt Anchor bushing include medium carbon steel with 0.42% Carbon and 93.98% Fe mixture of Nickel 1.91% Ni. Bolt Anchor Bushing has the characteristics of being resilient and corrosion resistant. The results of the microstructure on the outer, middle and inner surface of the bolt anchor bushing consist of ferrite and pearlite phases. The colour of ferrite phase is light and the dark is colour of pearlite phase. The ferrite phase is stable at low temperatures, solubility is limited and can be with Fe_3C or cementite. The pearlite phase is in ferrite and Fe_3C. In this study, the bushing consisted of Fe-Ni. It has the properties of high hardness, ductility, corrosion resistance and high temperature resistance.

1. Introduction

Bolt Anchor bushing is a machine element that has a loaded shaft so that rotation or alternating movements can take place smoothly and safely. The main function of the Bolt Anchor bushing is as a bearing between movable and immovable components. Bolt Anchor Bushing must have sturdy and strong properties for the shaft elements for the machine to work properly. If the bushing is not working properly, it will affect the entire system components and may not work properly. The brake assembly has bushings (rings) that attach to the arms of the main landing gear. The torque take-up slot is aligned with the torque pin. The torque take-up slot is aligned with the torque pin. Rotational friction occurs so that braking occurs. Main wheel brake assembly is shown in Figure 1, and bolt anchor is shown in description number 19 after Figure 1. Main landing gear instalation is shown in Figure 2.

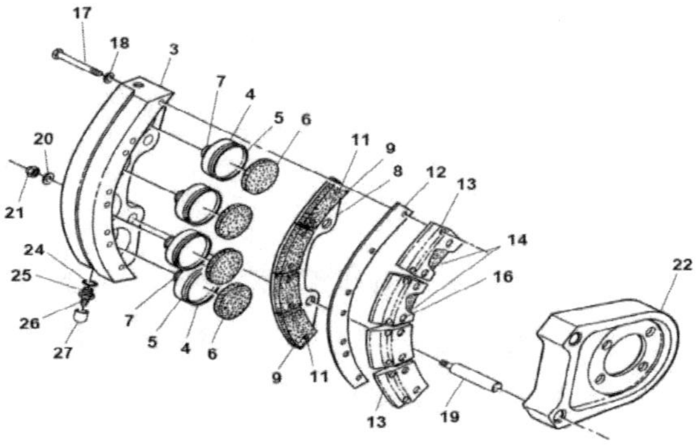

Figure 1. Main Wheel Brake Assembly
(Source: Cessna Company, 2019)

The description in Figure 1 is as follow:

1. Brake Assy Main Wheel
2. Brake Assy main Wheel
3. Silinder Brake
4. Piston Assy Brake
5. **5.** O-Ring
6. Insulator
7. Spring Friction
8. Plate Assy Pressure
9. Lining Brake Organic
10. Lining Brake Metalic
11. Pin Carrier
12. Shim Brake Plate
13. Back Plate Assy
14. Lining Brake Organic
15. Lining Brake Metalic
16. Pin Carrier
17. Bolt
18. Washer
19. **Bolt Anchor**
20. Washer
21. Nut
22. Torque Plate Assy Brake
23. Torque Plate Assy Brake
24. O-ring
25. Seat Bleeder
26. Screw Bleeder
27. Cap Bleeder

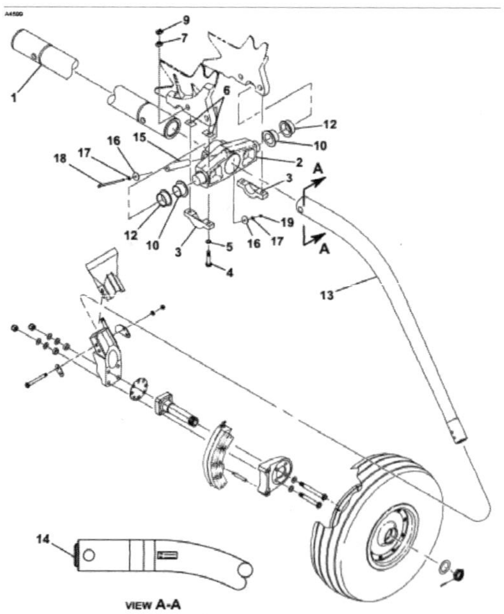

Figure 2. Main Landing Gear Instalation
(Source: Cessna Company, 2019)

The description in Figure 2 is as follow:

1. Spring Main Gear Center
2. Trunnion Assy main Gear
3. Cap main landing gear fitting
4. Bolt
5. Washer
6. Shim main Landing gear
7. Barrel Nut
8. Barrel Nut Alt
9. Retainer
10. Race main gear Tunnion
11. Rub strip gear Tunnion
12. Bearing main gear Tunnion
13. Spring main Gear
14. Plug Main Gear String
15. Pin Main Gear
16. Washer
17. Washer
18. Nut

2. Literature Review

Bushings made of SP2515, SCP5050 and SCP50094 tested at 364 ° C have good wear according to the tribometer [1]. Bolth anchor bushing for serving stand braking [2]. Bushings made of platinum-rhodium by hot isostatic pressing, it have environmentally properties and low manufacturing costs [3]. High strength, heat resistant, lightweight, corrosion resistant materials such as titanium, aluminum, composites are very suitable for aircraft materials [4].

Root the fuselage using rivet bolts with a finite element system to the installation process [5]. The joints on the fuselage consist of 2023-T3 aluminum alloy and the excellent bonding is 5x higher than the pure joint [6]. Copper alloy bushings having a high power capacity for deformation are used in aircraft engines [7]. Multi-axial axial compressor in modern aircraft engines and high heads [8]. Testing the bolt anchor bushing hardness using experimental theory and methods [9]. Rob bushing has difficulty controlling the dimensional accuracy of the material [10]. Bushing made of zirconium ceramic [11]. The finite element method and Catia program are very useful for connecting lugs to aircraft [12]. The MD 80 aircraft shows poor results because effect PH on the corrosion rate [13]. Method of energy dissipation in nail joints in plane structures by replacing parts of the hysteresis loop joint [14]. Bushing power is increased by reprocessing with billet extrusion. The effect of increasing the friction coefficient on the extrusion speed and billet heat will increase the billet breakdown and extrusion speed [15].

Ti6A14V titanium alloys are used in the aviation industry but there are drawbacks that titanium is difficult to cut but lightweight and highly corrosion resistant [16]. Research on aircraft maintenance must be carried out regularly and periodically so that safety is guaranteed [17]. Aluminum alloy 6061-T6 is suitable for aircraft bushings [18]. Bushing with a smooth surface quality consisting of Inconel Ti-6Al-4V material has high tool wear properties and is suitable for aircraft materials [19]. $AlSi_3O_4$ and Ti6Al4V and Inconel 718 materials are especially suitable for landing gear materials and brake systems because they have high wear resistance [20]. The brake lining characteristic of the Cessna 208B is a medium carbon steel alloy with 95.9% Fe and 0.33% C with a chromium alloy [21].

3. Method

3.1 Location: Polytechnic Manufacturing of Ceper, Klaten, Central Java.

IOP Conf. Series: Materials Science and Engineering 1158 (2021) 012020 doi:10.1088/1757-899X/1158/1/012020

3.2 Material: Bolt Anchor Bushing of Cessna Caravan 208B as in **Figure 3.**

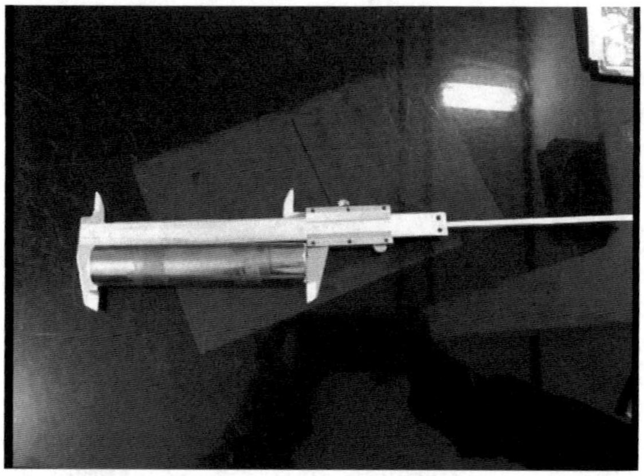

Figure 3. Bolt Anchor Bushing

3.3 Instrument :

Microscope optic for microstructure testing as in Figure 4 (a) and Optic Electron Spectroscopy as in Figure 4 (b), polishing machine in Figure 4 (c), and grinding machine in Figure 4 (d).

(a) (b) (c) (d)

Figure 4. Microscope optic for microstructure testing as in **Figure 4 (a)** and Optic Electron Spectroscopy as in **Figure 4 (b),** polishing machine in **Figure 4 (c)**, and grinding machine in **Figure 4 (d)**.

3.4 Method:

3.4.1. Test the chemical composition

The following steps for testing the composition are as follows: cleaning the specimen surface; placing the specimen on the test object; melt the test object; capturing light color; and record the chemical composition results.

3.4.2. Microstructure test

Prepare specimen, turn on machine in microstructure test, check in the magnifying lens; activate the microscope; scan an image at the desired magnification.

4. Results and Discuss

4.1 Results of Chemical Composition

The results of the characteristics of the Bolt anchor bushing in Cessna Caravan Type 208B as in Table 1 and the chemical element composition as in Figure 5.

Figure 5. Chemical Composition Testing

Table 1. Chemical Composition

Element	Chemical composition (%)
C	0.42994
Si	1.4876
Mn	0.71905
P	0.01894
S	0.00715
Cr	0.77823
Mo	0.43010
Ni	1.9188
Al	0.04225
Co	0.01971
Cu	0.04487
Nb	0.01975
Ti	0.01773
V	0.07847
W	0
Pb	0.0678
Fe	93.98243%

(Source: Polman Ceper, Klaten, Jawa Tengah)

The function of element in bolt anchor bushing in Cessna caravan Type 208B:

a. Iron (Fe)

Iron is the dominan element in the bolt anchor bushing. The iron element contained in the bolt anchor bushing formation is 93.98243%.

b. Nickel (Ni)

Ni has tough properties and is resistant to corrosion (rust), is heat resistant up to 1200°C but does not have high hardness. The bolt anchor bushing material in this study contained 1.9188% nickel elements.

c. Silicon (Si)

This bolt anchor bushing material contains 1.4876% silicon, which is ductile and heat resistant.

d. Chromium (Cr)

This element is used as a protective and friction resistant steel surface. This Cr is shiny, hard, brittle, and corrosion resistant. The chromium contained in this bolt anchor bushing material is 0.77823%.

e. Manganese (Mn)

This bushing material containing 0.71905% manganese has crack resistance.

f. Molybdenum (Mo)

Molybdenum contained in the bolt anchor bushing material is 0.43010% which has friction and heat resistance properties.

g. Carbon (C)

Carbon function in friction resistant steel. Contains 0.492% Carbon.

Bolt author Bushing is a component contained in an airplane in the form of a cylindrical iron tube in which there is rubber. The bolt author bushing functions as a cushion between moving and immovable components so that the functions of the aircraft components can carry out their functions properly and safely. The results of the chemical composition in the Bolt author bushing include medium carbon steel with 0.42% Carbon and 93.98% Fe, a mixture of Nickel 1.91% Ni. Bolt anchor bushing has the characteristics of ductility and corrosion resistance. The microstructure results on the outer, middle and inner surfaces of the bushings consisting of ferrite and pearlite. The light is colour of ferrite phase and the dark is colour of pearlite. The ferrite phase is stable at low temperatures, solubility is limited and can be with Fe_3C or cementite. The pearlite phase is in ferrite and Fe_3C. In this study, the bolt anchor bushing consisted of Fe-Ni. This bolt author bushing has the properties of high hardness, high ductility, corrosion resistance, and resistance to high temperatures.

4.2 Results of Microstructure

Three images of microstructure results on the Cessna Caravan 208B bolt anchor bushing, the outer, middle, and inner surfaces have ferrite and perlite phases. The light is colour of ferrite phase and the dark is colour of pearlite. It as in Figure 6. Microstructure 100x magnification; Microstructure 50x magnification as in Figure 7 and Microstructure 200x magnification as in Figure 8.

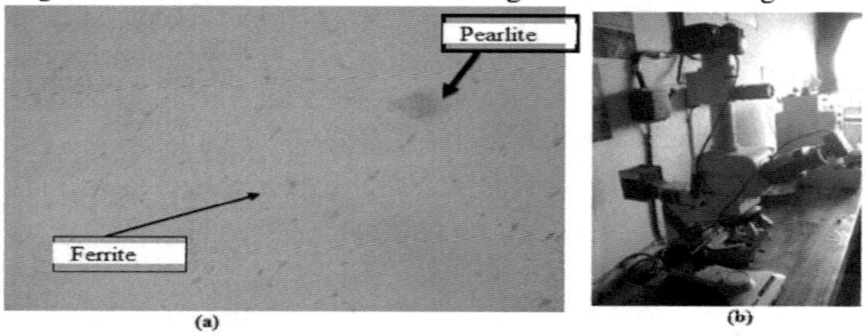

Figure 6. Microstructure 100x magnification with microscope optic (Source: Polman Ceper, Klaten)

Figure 7. Microstructure 50x magnification with microscope optic (Source: Polman Ceper, Klaten)

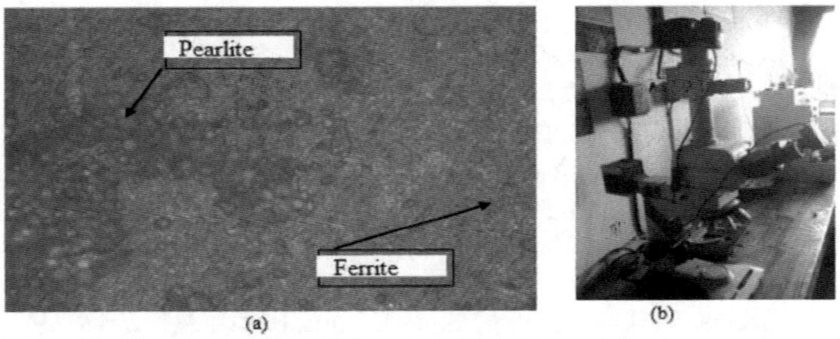

Figure 8. Microstructure 200x magnification with microscope optic
(Source: Polman Ceper, Klaten)

This corresponds to the carbon content of the material which is 0.42994%, so the element is classified as medium carbon steel. In addition to the carbon element, there is also the largest element, namely the element iron (Fe) 93.98243%, where the iron element has good hardness and good ductility. Silicon (Si) 0.14876%, where the silicon content has elastic properties and has high ductility, besides silicon, it also adds hardness and sharpness to steel. In addition to the silicon element there is also an element of manganese (Mn) 0.71095%, the content of manganese has properties that are resistant to friction and pressure resistance (impact load).

Nickel (Ni) 1.9188%, nickel has properties that are ductile and resistant to chemicals to deal with serious corrosion (rust) but does not have high hardness, nickel is an element that is mixed into steel to overcome damage at high temperatures. Molybdenum (Mo) 0.43010%, has heat resistance properties making it suitable for hot tool steel, the maximum mixing limit of this element is 7% and also serves to neutralize the hardness of tungsten. Molybdenum is an additional element of maximum steel ductility. Chromium (Cr) 0.77823%, this element is used as a protective and friction resistant steel surface. Steel is shiny, hard and brittle and resistant to corrosion. Perlite in medium carbon steel is a mixture of ferrite and cementite. Perlite is an alloy of ferrite and cementite. Perlite has high hardness when the iron element is high. Perlite is shiny, ferrite is light in color, and cementite is dark in color. High carbon steel is an alloy of perlite and cementite while mild carbon steel is a blend of pearlite and ferrite.

5. Conclusion

Bolt author bushing is a component contained in an airplane in the form of a cylindrical iron tube in which there is rubber. The bolt author bushing functions as a cushion between moving and immovable components so that the functions of the aircraft components can carry out their functions properly and safely. The results of the chemical composition in the Bolt author bushing include medium carbon steel with 0.42% Carbon and 93.98% Fe, a mixture of Nickel 1.91% Ni. Bushing has the characteristics of ductility and corrosion resistance. The microstructure results on the outer, middle and inner surfaces of the bushings consisting of ferrite and pearlite. The ferrite phase is light in color and the pearlite phase is dark. The ferrite phase is stable at low temperatures, solubility is limited and it can be with Fe_3C or cementite. The pearlite phase is in ferrite and Fe_3C. In this study, the bushing consisted of Fe-Ni. This bolt author bushing has the properties of high hardness, high ductility, corrosion resistance, and resistance to high temperatures.

Acknowledgements
The author would like to Yayasan Citra Dirgantara and STTKD Yogyakarta in this opportunity.

References

[1] Chen, S., Yin, N., Yu, Q., and Zhang, Z. 2019. A novel tribometer for investigating bushing wear. *Wear*, **430**, pp.263-271.
[2] Tuan, C. Y., and Dass, W., C. 2016. Expedient airfield runway repair using folded fibreglass mat. *International Journal of Pavement Engineering*. **174**, pp. 283-299.

[3] Caretto, F., Laera, A. M., Di Nuzzo, F., Iovino, R., Di Benedetto, F., Pesce, E., ... and Tapfer, L. 2016. Molybdenum disilicide-silicon nitride bushing nozzles tailor-made for basalt fibers production. *Ceramics International*, **42**(10), pp. 11844-11850.

[4] Ilg, J., Müller, T., and Kuhn, S. 2017. Dry Machining of Multilayer Composite Materials in Aircraft Construction. *Lightweight Design worldwide*, **10**(6), pp. 42-47.

[5] Chernykh, E. V., Zaidès, C. A., and Kudryavtsev, A. A. 2019. Peculiarities of making bolt and rivet connections with interference when assembling the aircraft MS-21. In *IOP Conference Series: Materials Science and Engineering,* **632**(1), p. 012090.

[6] Pitta, S., de la Mora Carles, V., Roure, F., Crespo, D., and Rojas, J. I. 2018. On the static strength of aluminium and carbon fibre aircraft lap joint repairs. *Composite Structures*, **201**, pp. 276-290.

[7] SHU, C., Fan, W., Wang, Z., and Zhang, J. 2015. Study on Optimization of Plastic Forming Process for Copper Alloy Rod Bushing. *Hot Working Technology*, **09**.

[8] Boyko, L., and Dyomin, A. 2018. Numerical study of flows in axial compressors of aircraft gas-turbine engines. *Восточно-Европейский журнал передовых технологий*, **4**(8), pp. 40-49.

[9] Lu, S., Zhang, B., Wang, Z., and Hu, X. 2017. Experimental Study on Influence of Spinning Forming Process Parameters on Quality and Performance of Connecting Rod Bushing. *Hot Working Technology*, **03**.

[10] Lv, W., Fan, W. X., Wang, Y., and Li, J. H. 2016. Dimensional accuracy analysis of spinning process of connecting rob bushing. *Journal of Plasticity Engineering*, **03**.

[11] Alisin, V. V., and Roshchin, M. N. 2019. Research of Friction Type Bearing with Zirconium Ceramics Bushing. In *2019 International Multi-Conference on Industrial Engineering and Modern Technologies (FarEastCon)* (pp. 1-4).

[12] Phadnis, P., Anilkumar, G., and Madhusudhan, T. 2015. Review on fatigue and damage tolerance evaluation of aircraft lug-joint. *International Research Journal of Engineering and Technology (IRJET)*, **502**.

[13] Suhartaya, N. U. C., and Rochmadi, R. 2020. The effect of pH upon corrosion rate of aircraft component type MD 80 on fuselage part. In *IOP Conference Series: Materials Science and Engineering,* **732**(1), p. 012089.

[14] Tachi, T., Machida, S., Okada, T., and Suzuki, S. 2017. Simple Energy Dissipation Model of Riveted Lap Joint in Aircraft Structure. *Transactions of The Japan Society for Aeronautical and Space Sciences*, **60**(6), pp. 341-346.

[15] Wenxin, F., Cuncun, C., Hualong, Y., and Wenbo, Y. 2016. Simulation analysis on damage of connecting rod bushing in warm extrusion based on finite element. *Forging & Stamping Technology*, **08**.

[16] Li, L., Xue, H., and Wu, P. 2018. Experimental study on the axis line deflection of Ti6A14V titanium alloy in gun-drilling process. In *IOP Conference Series: Materials Science and Engineering,* **301**(1), pp. 012012.

[17] Nathanael, D., Tsagkas, V., and Marmaras, N. 2016. Trade-offs among factors shaping operators decision-making: the case of aircraft maintenance technicians. *Cognition, Technology & Work*, **18**(4), pp. 807-820.

[18] Phadnis, P., Anilkumar, G., and Madhusudhan, T. 2015. Review on fatigue and damage tolerance evaluation of aircraft lug-joint. *International Research Journal of Engineering and Technology (IRJET)*, **502**.

[19] Dehghan, S., and Baharudin, B. T. 2019. Friction drilling of difficult-to-machine materials: workpiece microstructural alterations and tool wear. *Metals*, **9**(9), pp. 945.

[20] Dehghan, S., Ismail, M. I. S., Ariffin, M. K. A., and Baharudin, B. T. H. T. 2019. Measurement and analysis of thrust force and torque in friction drilling of difficult-to-machine materials. *The International Journal of Advanced Manufacturing Technology*, **105**(7), pp. 2749-2769.

[21] Suroso, I. 2020. Characteristics of Brake Lining Cessna Caravan Type 208B. In *IOP Conference Series: Materials Science and Engineering*, **722**(1), p. 012027.

Development of Fly Ash Coal/TiO_2 Pored Composite Materials in The Making of Ceramic Membrane for Water Treatment Process

Eny Apriyanti[1,*], Heru Susanto[2], and I Nyoman Widiasa[3]
[1,2,3]Department of Chemical Engineering, University of Diponegoro, Jl. Prof. Sudarto, Semarang, Central java, Indonesia
[2,3]Membrane Research Center (MeR-C), University of Diponegoro, Jl. Prof. Sudarto, Semarang, Central java, Indonesia

Email: *enyapriyanti66@gmail.com

Abstract. The ceramic membrane is a synthetic membrane, also called an inorganic membrane. The objectives of this study are to evaluate the characteristics of fly ash ceramic membrane and their performance for water treatment. The ceramic membrane fabrication process uses a printing process with a pressure of 1 atm with a burning temperature of 1100°C for 7 hours. The stages of this research include 1) composition optimization experiments with variations in the composition of fly ash and kaolin clay and temperature, 2) Characterization of ceramic membranes obtained by X-ray diffraction (XRD) and scanning electron microscope (SEM) tests. Optimization is carried out by varying the composition using a ratio of fly ash kaolin clay MB1: 40%: 20 %: 30%; MB2: 30%: 30%: 30%; MB3: 20%: 40%: 30% with a combustion temperature of 1100°C with a combustion process time of 7 hours. From the research results obtained, the porosity test results MB1 porosity 42%; The porosity of MB2 is 44%; The porosity MB3 is 50.22% from the porosity test results showing the higher composition of coal fly ash, the smaller porosity obtained. This phenomenon occurs during the combustion or sintering process to minimize the membrane pores. The structure of the membrane pores can be seen by the characterization test using SEM. SEM test results show that the MB1 ceramic membrane has delicate pores so that it has the potential to be applied. The experimental results showed that the utilization of fly ash for ceramic membrane material was promising for water treatment due to its excellent properties and filtration performance.

1. Introduction

Ceramic membranes are inorganic membranes currently being developed into composite membranes ceramic and have many advantages compared to other membranes. Ceramic membranes have resistance to heat, acids, and bases [1]. The development of composite ceramic materials is increasing due to their application. The main ingredients of ceramic membranes such as alumina, titania, zirconia, silica, fly ash, clay; for the use of ceramic membrane materials, it is necessary to study natural alternative materials such as clay, which is waste from the PLTU, the two materials have the potential to be used as a mixture for making composite ceramic membranes [2]. While pore-forming materials other than inorganic materials such as carbon and calcium carbonate and organic materials such as fly ash, in principle, pore-forming materials are flammable materials [3]. Based on several researchers' studies, this study uses fly ash as the raw material and as a pore-forming material needed in the manufacture of ceramic membranes. The performance of ceramic membranes for the separation process is usually expressed in terms of flux.

Ceramic membrane flux is related to porosity; the more-high porosity will have better filtering capabilities [4]. The use of coal fly ash has been carried out in many previous studies because it is one of the non-metallic materials of ceramic in the form of oxygen, carbon, nitrogen and silicon bonding compounds, which are strong and hard and corrosion resistant. Another property of this non–metallic materials is that it has a low density and high melting point, coarse particles and contributes greatly to the mechanical properties of the hardness of the material because the material is not easy to soften and is resistant to penetration on its surface [5]. Ceramic membranes made from a mixture of fly ash and kaolin can process river into clean water that meets clean water standards under Regulation of Indonesia Health Ministry No. 416/Menkes/Per/XI/1990 [6]. Several studies have been conducted [7] , [8] examining the coal fly ash formula in making ceramic membranes that can increase the flux value. The resulting flux value for river water is 469.721/m^2.hours at a pressure of 1.00 bar.

Research using porous composite coal fly ash materials has been carried out by several previous researchers, including the dry pressing method, which was carried out using specific pressures. After that, the composite material mixture is put in the mold and pressed. After that, the mold is entered into the furnace for the sintering process [3]. The result shows that the use of clay-based materials can be utilized for the manufacture of ceramic membranes with a pure alumina content of 70.6% SiO_2, having a specific area of 98m^2/g and particle size of 0.3-1 μm or 300 – 1000 nm. The study of coal ash incorporation into montmorillonite matrix and the optimization of composition and combustion temperature in ceramic membrane fabrication has been reported [9]. The use of coal ash as a ceramic membrane material as peat water treatment with the influence of the weight composition of coal ash and clay can treat peat water optimally at temperature burning 900°C [10]. Besides, coal ash also contains other minor minerals such as magnesium, sulfur, sodium, potassium, and carbon [5]. The development of porous ceramic materials is increasing because of its applications covering all fields, especially those that prioritize high temperatures. One of the ceramic materials currently being developed is coal fly ash as a raw material in the ceramic materials currently being developed is coal fly ash as a raw material in the manufacture of ceramic membranes due to the abundant availability of materials to produce ceramic membranes with lower prices, ceramic membrane provides many advantages over polymer membranes. After all, ceramic membranes provide many advantages over polymer membranes because ceramic membranes are stable at high temperatures and have good mechanical strength. They have properties that do not expand quickly in water, easily form a suspension to coat the membrane support [11]. Based on the background description, it is necessary to study the ability of ceramic membranes using the microfiltration method for the clean water purification process. As an effort to develop the clean water treatment process, a preparation process is carried out to determine the composition of the material and the technique for making coal ash-based tubular separated. Ceramic membranes so that they can be used in clean water purification with membrane technology is a separation process using a membrane where the feed to be separated is contacted with the membrane surface, while the lower pressure permeate. Membrane characteristic tests include XRD, SEM, Fourier Transform Infra-Red (FTIR), permeability analysis. This study aims to produce ceramic membranes or porous materials based on coal ash, kaolin clay and alumina for water treatment, the specific objective of this study is to examine the optimal composition of alumina and kaolin which affects the preparation of coal ash-based tubular ceramic membranes.

The performance of the ceramic membrane for the separation process is usually expressed in terms of flux. The flux of the ceramic membrane is related to porosity. The higher the porosity, the better the filterability. In general, the manufacture of TiO_2 composite membranes includes i) the manufacture of a porous membrane for the support layer, which is usually made by phase inversion techniques, and ii) Selective layer deposition on the surface of the support layer [12]. It is hoped that the development of composite materials into ceramic membranes can be used for clean water treatment, resulting in meeting the clean water quality standard set in Indonesia under the Minister of Health Decree No.907/MENKES/SK/VII/2002.

2. Materials and Methods

The research process was carried out in several stages: 1) Preparation of coal fly ash in the form of solid waste, 2) Making the support layer by optimizing the composition of the mixing of the various mass of fly ash, alumina, kaolin as a support layer and additives, pugging, aging, printing and calcination, 3) Characterization of the support layer and active layer consists of Scanning electron microscopy test.

2.1. Raw materials preparation

The materials used for the manufacture of fly ash ceramic membranes, kaolin, alumina, and additives as auxiliary materials in the manufacture of ceramic membranes are preparing raw materials with a mortar and stamper after they are smooth filtered using a 200-mesh shaker. Filtering here is done so that the size of the ceramic membrane constituent materials can be mixed homogeneously to not interfere with the next process. After the ingredients are filtered, then they are composed according to the predetermined research variables.

2.2. Fabrication of supporting layer

The first process carried out is material preparation. One of the materials prepared is to fly ash. We sift fly ash using a shaker until the size is homogeneous to 200 mesh. The experiment was started by mixing the main support layer ingredients, fly ash, kaolin, and alumina, and making a mixture of additives in the carboxymethyl cellulose, polyethylene glycol, sodium citrate, and $MgSO_4$, respectively, according to the determined composition. After that, the main ingredients mixture and the additive mixture, and the additive mixture are mixed for 15 minutes. The next step is pugging the mixture that has been formed for 30 minutes to form a homogeneous and elastic paste. The paste is then left at room temperature and kept away from direct sunlight for 30 minutes (aging process), and then the paste is printed into flat sheets. The results of these prints were then dried at a temperature of 250°C for one hour to remove their organic content, after which the support membrane was calcined at 1100°C for 7 hours. The process of making TiO_2 composite ceramic membranes using several core tools and research support tools, and a series of research tools is shown in Figure 1.

2.3. Ceramic membrane characterizations

The characterization of ceramic membranes was carried out to determine the mechanical strength of membranes through permeability and porosity test to water. Besides that, ceramic membranes were measured pores and their pore distribution using Scanning Electron Microscopy (SEM).

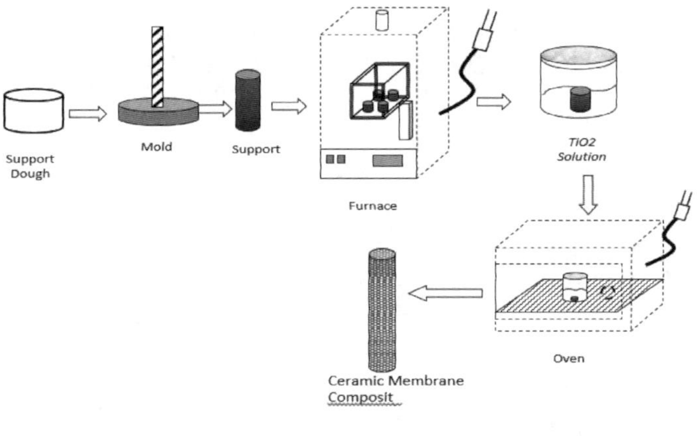

Figure 1. Schematic diagram of TiO_2 composite ceramic membrane fabrication

Figure 1 shows the method of making porous ceramic membranes based on coal ash, an overview of the process of this study using dip coating and dry pressing in which all the ingredients are mixed and mixed to obtains a homogeneous mixture after the mixture is put in the mold and done with an emphasis on certain pressure.

3. Results and discussions

In this study, the variables used to make the support layer are the composition of the material. The materials for the support layer are varied, as can be seen in Table 1. It shows the composition of the materials and additives used to make composite ceramic membranes with different variations in composition I, composition II and composition III.

Table 1. Variation of Support layer composition

Composition I	Composition II	Composition III
Fly Ash: 40 %wt Alumina: 20 % wt Kaolin: 30 %wt Additive : 10 %wt	Fly Ash: 30 %wt Alumina: 30 % wt Kaolin: 30 %wt Additive : 10 %wt	Fly Ash: 20 %wt Alumina: 40 % wt Kaolin: 30 %wt Additive: 10 %wt
Deionized Water : 30 %wt From total Weight of ra material	30 %wt From total Weight of raw material	30 %wt From total Weight of raw material

3.1. *Porosity test*

The porosity test is carried out to determine the amount of substance or components absorbed by the membrane. The porosity test is usually carried out on the water to see the amount of water that can be absorbed by the membrane. The method used to perform the porosity test is by immersing the membrane in water for 24 hours at room temperature, then the membrane is weighed, after which the membrane is dried in an oven at 60°C for 48 hours it is scorched then it is weighed. This test is done by immersing the ceramic membrane in boiling water, then considering it. The porosity test result can be seen in Table 2.

Table 2. Ceramic Membrane Porosity Test Result

Membrane	Fly ash material Composition	Membrane weight			Porosity (%)	Average
		Dry (gram)	Wet (gram)	Difference (gram)		
MB1	40%	42.79	60.98	18.19	42.51	
		42.65	61.04	18.38	42.90	42.82
		41.97	60.46	18.07	43.05	
MB2	30%	41.12	60.92	19.80	48.16	
		39.14	53.19	14.05	35.90	43.56
		42.32	62.05	19.73	46.62	
MB3	20%	37.21	55.40	18.19	48.88	
		36.94	56.23	19.29	52.22	50.22
		38.02	56.87	18.85	49.58	

Table 2 shows the result of the porosity test. The greater the fly ash composition, the smaller the porosity value this happens because of the influence of the combustion or sintering process to reduce the membrane porosity.

3.2. *Permeability test*

The result of the permeability test or the result of the flux test show the relationship or rejection and flux concentrations to produce the optimum work at a concentration of 2%. Figure 2 shows the magnitude of the permeability to water from film made from various concentrations of TiO_2 solution in nitric acid, which significantly affects. The higher the TiO_2 concentration caused the lower film permeability. This result is because of the higher the TiO_2 concentration, the higher the solution's viscosity. The resistance caused by the film is also higher, and the volumetric rate of water becomes smaller. At concentrations of TiO_2 above 3%, TiO_2 cannot dissolve entirely, so the resulting film is not homogeneous and porous. This phenomenon will result in an increase in flow rate, the effect of the permeability value on the increase in solution concentration, and the characteristic of TiO_2 based on the molecular weight of the rejection.

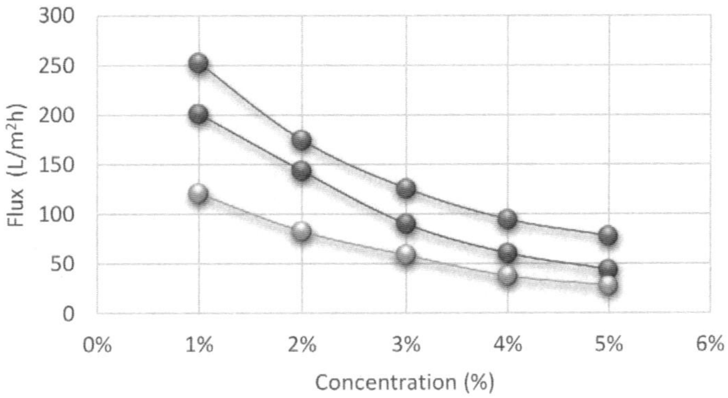

Figure 2. Membrane permeability test result

3.3. SEM (Scanning Electron Microscopy) Test

For the observation of the material using the SEM test in Figure 3, it can be concluded that the presence of air trapped in the porous material when sintering takes place Al_2O_3 will experience more expansion than SiO_2; in this case, it is influenced by membrane casting pressure, the greater membrane pore size, this is influenced by the combustion temperature or the calcination process carried out at a temperature of 1100°C.

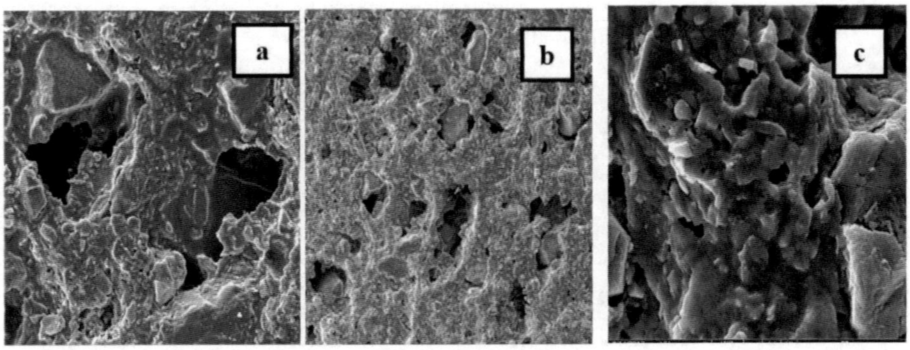

Figure 3. SEM images of ceramic composite membranes: (a) MB1-40 wt.% fly ash composition, (b) MB2-30 wt.% fly ash composition, and (c) MB3-20 wt.% fly ash composition

From the SEM test result, it can be seen that there are white spots on the surface of the membrane morphology, which indicates a dispersion factor between additive particles and polymer particles. In this study, the SEM result showed that the addition of PEG (Polyethylene Glycol) additives increased the pore size to a maximum at a concentration of 2%. Subsequent addition to reducing pore size was due to additive agglomeration so that it could fill the membrane pores up to 10% by weight of Al_2O_3 indicates the maximum number of membrane pores.

4. Conclusion

Based on coal fly ash concentrate and kaolin clay as raw material using the dip-coating method with TiO_2, the resulting membrane structure is more robust and not easily cracked at operating conditions P= 1 bar and T calcination = 1100°C with optimum concentration was 2 wt%. The higher composition of coal fly ash produced lower porosity where the highest porosity was achieved by MB3 of 50.22%. The higher the TiO_2 concentration exhibits, the lower film permeability. However, the concentration of TiO_2 more than 3 wt.% resulting inhomogeneous and porous membrane. The dominant crystal structure morphology of the coal ash membrane is the Al_2O_3 as proven by SEM images. Overall, the result of the ceramic composite membrane analysis test shows that coal ash can be used as raw material for the manufacture of ceramic membranes which produces a strong membrane and has good porosity so that it can be applied for clean water treatment.

References

[1] Hristov, P., Yoleva, A., Djambazov, S., Chukovska, I., and Dimitrov, D. 2012. Preparation and characterization of porous ceramic membranes for micro-filtration from natural zeolite. *Journal of the University of Chemical Technology and Metallurgy*, **47**(4), pp.476-480.

[2] Diana, S., Zaharani, L., and Fona, Z. 2019. Pemanfaatan Fly Ash dan Clay dalam Pembuatan Membran Keramik dengan Penambahan PVA Sebagai Perekat untuk Merejeksi TSS pada Air Sungai. *In Prosiding Seminar Nasional Politeknik Negeri Lhokseumawe*, **2**(1), pp.91–95.

[3] Iman, R. 2017. Manufacture and Characterzation of Ceramic Membrane Variation of Rice Flour as *An Additive for The Microfiltration Process J. Sains Terapan Kim.* **11**(2), pp.52–60.

[4] Chasri, N. 2015. The Utilization of Coal Fly Ash as Ceramic Membranes for the Unit of Peat Water Treatment J. Dinamika Kim. **26**(2), pp.95–105.

[5] Fang, J., Qin, G., Wei, W., and Zhao, X. 2011. Preparation and characterization of tubular supported ceramic microfiltration membranes from fly ash Sep. Pur. Technol. **80**(3), pp. 585–591.

[6] Agmalini, S., Lingga, N. N., and Nasir, S. 2017. Peningkatan Kualitas Air Rawa Menggunakan Membran Keramik Berbahan Tanah Liat dan Abu Terbang Batubara J. Tek. Kim. **19**(2), pp. 59–68.

[7] Zhang, Y., Liu, F., Lu, Y., Zhao, L., and Song, L. 2013. Investigation of phosphorylated TiO2–SiO2 particles/polysulfone composite membrane for wastewater treatment Desalination, **324**, pp. 118–126.

[8] Suresh K 2020 Recycling of fly ash for development of value added products Mater. Today: Proc., S2214785320373788.

[9] Jedidi, I. 2009. New ceramic microfiltration membranes from mineral coal fly ash," Arabian J. Chem. **2**(1), pp. 31–39.

[10] Zhou, W., Zhang, L., Wu, P., Cai, Y., Zhao, X., and Yao, C. 2019. Study on Permeability Stability of Sand-Based Microporous Ceramic Filter Membrane Materials, **12**(13), pp. 2161.

[11] Dong, Y. 2010. Recycling of fly ash for preparing porous mullite membrane supports with titania addition J. Hazard. Mater. **180**(1–3), pp. 173–180.

[12] Mulder, M. 1996. Basic Principles of Membrane Technology. Dordrecht: Springer Netherlands.

AUTHOR INDEX